New Trains for New Railways

Conference Organizing Committee

A Baker (Chairman)
Angel Train Contracts

D Razdan
ADtranz

K Majoribanks
Angel Train Contracts

J Vine
Halcrow Transmark

P Watts
GEC Alsthom

IMechE
Conference Transactions

I MECH E

Conference on

New Trains for New Railways

20–21 May 1998
York Viking Moat House, UK

Organized by the Railway Division of the
Institution of Mechanical Engineers (IMechE)

Sponsored by

KVÆRNER ATC AMEC Rail AEA TECHNOLOGY RAIL

Co-sponsored by the

Railway Industry Association
Institution of Railway Signal Engineers
Institution of Civil Engineers (ICE)
Institution of Electrical Engineers (IEE)
Railway Forum
Office of Passenger Rail Franchising (OPRAF)

IMechE Conference Transactions 1998–7

Professional Engineering Publishing

Published by Professional Engineering Publishing Limited for
The Institution of Mechanical Engineers, Bury St Edmunds and London, UK.

First Published 1998

ISSN 1356–1448
ISBN 1 86058 146 3

A CIP catalogue record for this book is available from the British Library.

Printed by The Ipswich Book Company, Suffolk, UK.

Contents

Related Titles of Interest

Title	Editor/Author	ISBN
Train Maintenance Tomorrow and Beyond	IMechE Conference 1997–1	1 86058 095 5
Better Journey Time – Better Business	IMechE Conference 1996–8	0 85298 997 0
Implementing Rail Projects	IMechE Conference 1995–1	0 85298 948 2
Railway Traction and Braking (Railtech '96)	IMechE Seminar 1996–19	1 86058 018 1
Design, Reliability and Maintenance for Railways (Railtech '96)	IMechE Seminar 1996–18	1 86058 017 3
Railway Rolling Stock (Railtech '96)	IMechE Seminar 1996–17	1 86058 016 5
Railway Engineering, System and Safety (Railtech '96)	IMechE Seminar 1996–16	1 86058 015 7

For the full range of titles published by Professional Engineering Publishing contact:

Sales Department
Professional Engineering Publishing Limited
Northgate Avenue
Bury St Edmunds
Suffolk
IP32 6BW
UK

Tel: 01284 724384
Fax: 01284 718692

C547/001/98

New trains for a new railway – a global perspective

P METCALF CEng, MIMechE
Halcrow Transmark, Derby, UK
D McCABE
Walkers Limited, Queensland, Australia

SYNOPSIS

As the momentum in new train investment gathers pace in the United Kingdom, world-wide investment and development in new railways has been ongoing for some time. This paper describes investment in railways in Japan, Australia and Romania and ultimately questions whether lessons learnt will have any impact
either directly or indirectly upon the United Kingdom market.

1. INTRODUCTION

As we approach the new millennium, the growth in railway investment in the United Kingdom gathers pace at a rate which is perhaps unprecedented in recent times.

However, world-wide in the field of Traction and Rolling Stock, investment has been ongoing for some time.

Against this background the issues raised in this paper are focused around both commercial and technical aspects of new/refurbished train procurement.

Three examples of investment around the world have been chosen which could be considered to give a broad picture of investment in new/refurbished trains, and as the title of this paper states, will provide a truly global perspective.

From Japan the Railway of Japan Central, in particular the Tokaido Shinkansen.

From Queensland Australia, the new tilting train which will be entering service in September 1998, and finally from Romania the investment which is ongoing in refurbished Traction and Rolling Stock.

From these three examples the paper will be able to address some key issues such as :

a) What were the primary levers for investment?
b) What was the degree of technical innovation ?

c) What were the technical risks?
d) Was there technology transfer?

Finally the paper will conclude by providing an independent view as to what significance the issues raised could have on new train investment in the United Kingdom.

2. JAPAN RAILWAY COMPANY (JR CENTRAL)

2.1 Background

In the global world of railway operation, JR Central is perhaps one of the best examples of a reliable, technically advanced railway, successfully married to the demanding commercial objectives of a major business.

The Company was established in 1987, following both the privatisation and subsequent division of Japanese National Railways (JNR). It inherited the Tokaido Shinkansen and part of the Conventional Line network. In total some 1946 km of line.

It's principal business objective is focused upon preserving and developing further the volume of passenger traffic which it carries.

The Tokaido Shinkansen is the jewel in the Company crown, and has the stature of the world's busiest high-speed rail line with up to eleven departures per hour at peak times, linking the major business centres of Tokyo, Nagoya and Osaka.

Japan Central also operates the Conventional Line network which relative to this paper is a secondary aspect of the business, but none the less just as interesting, due to the technology transfer aspects which will be referenced at a later stage. The Conventional Lines also link into the Shinkansen, at station interchanges, so maximising the operation of the total network, a practice copied by airline carriers with hub and spoke operation.

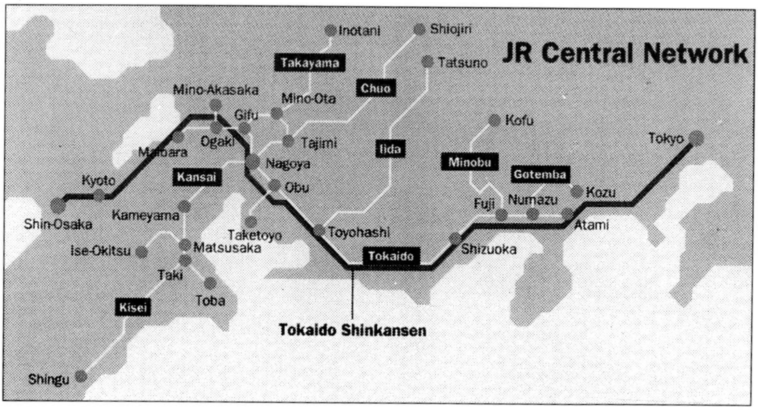

Figure 1 - Operating Network Of Japan Central

In the context of this conference, the achievements and present status of this Railway operation perhaps bears a striking resemblance to the position which Train Operators/Owners in the United Kingdom aspire to.

Salient characteristics being:

- Trains which run to time - Average delay per train is less than 1 minute
- High vehicle/component reliability
- High sustainable levels of passenger volume growth
- Healthy profit stream
- Continuing development of new products (new trains)

2.2 The Case For Investment

The megalopolis of Tokaido accounts for 46% of the Japanese Gross Domestic Product, and 36% of the Country's population, whilst it only accounts for 10% of Japan's total land mass.

As the Japanese economy expanded during the 1950's the existing railway system had reached it's capacity in terms of speed of operation, performance etc. Subsequently a strategically challenging decision was taken by the President of Japan National Railways (JNR) to build a complete new line between Tokyo and Osaka, based on the forecast at the time (1957) that traffic demand would double by the mid 1970's.

This line was opened in 1964 in time for the Olympic games, and heralded the introduction of the world famous Bullet Train (Shinkansen). Although the project was completed on time the cost overrun was significant, therefore the President and Vice President of JNR felt compelled to resign. The journey time on the new railway known as the Shinkansen was a remarkable three hours for a distance between the two cities of 515 Km.

Passenger volumes for the new service in 1964 averaged 60,000 per day. Today the average daily passenger volume is 368,000. A case of success breeding success!

In total since its introduction the Tokaido Shinkansen has carried some 3 billion passengers. The strong steady growth in passenger volume has seen the cities along the route generate high levels of business. The railway, and this one in particular, is viewed as the critical factor in supporting both the social, and importantly the economic, development of Japan. Thus the bedrock for the investment case is founded for new trains with new technology. The results do speak for themselves with a sustainable increase in passenger volume coincident with an increase in GDP, see Figure 2, in other words the Railway stimulating the economy.

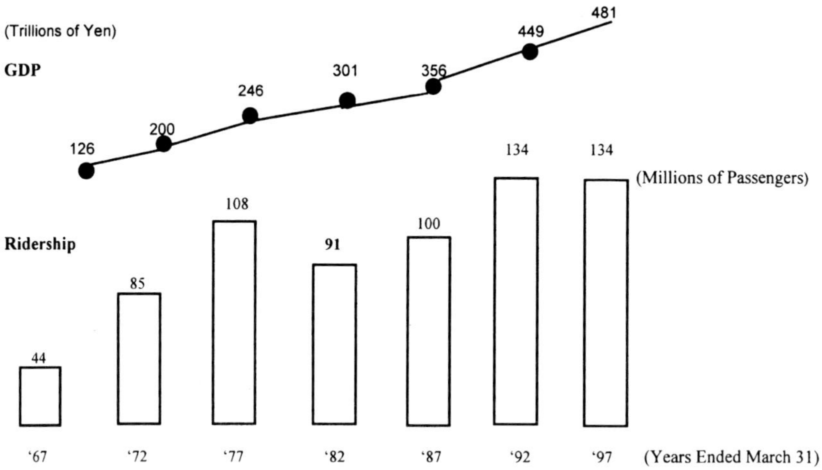

Note: GDP is expressed in 1990 figures.

Figure 2 -Japanese GDP Compared to Passenger Volume (1)

This view may well be anathema to the historical approach in the United Kingdom which has generally been that investment in new trains/new technology will only occur when the economic outlook is vibrant. What the Shinkansen epitomises is that it can contribute towards a vibrant economy.

The market share of the Tokaido Shinkansen compared to it's main competitor, namely the airlines, is that the railway presently has 84% of the market. Over other Shinkansen routes outside the Japan Central network the airlines have a much stronger market share, 48% between Tokyo and Hiroshima (821 km) and 88% between Tokyo and Fukuoka (1069 km) [1]. The primary reason being the journey time, with the airlines clearly having the competitive edge over the longer distances.

For Japan Central the need to maintain the market share is paramount and is demonstrated by several key initiatives, not least of which are the development of the Series 700 Shinkansen and the Super Conducting Maglev, both of which will be discussed later in this paper.

2.3 Degree Of Technical Innovation

Underpinning the business success of the Shinkansen has been the pace of technical developments relative to both the new trains and the infrastructure. Evolution rather than revolution are the watchwords. Nowhere is this better personified than with the Shinkansen Rolling stock.

Figure 3 - Three Generations Of Shinkansen Passenger Rolling Stock

There are presently three generations of these trains operating today on the Tokaido Shinkansen for J.R. Central.

Series 0 The original bullet train. Introduced in 1964 this train has the following basic features:

• Maximum speed	220 km/h	[1]
• Passenger capacity	1391	
• Weight/train set	970 ton	
• Vehicles/set	16 motored	
• Body material	Steel	
• Traction motors	DC series wound (64/trainset)	
• Traction control	Low-voltage tap switching step control	
• Number of pantographs	8/train set	
• Brake control system	Rheostatic brake with electro-magnetic straight air brake	
• Bogie	Bolster type with end beams	

Series 100 Introduced in 1985 the train has the following basic features:

• Maximum speed	220 km/h
• Passenger capacity	1321
• Weight/train set	925 ton
• Vehicles/set	12 motored/4 trailer
• Body material	Steel
• Traction motor	DC series wound, 48 motors/train set

- Number of pantographs 3 per train set
- Traction control Thyristor continuous phase control
- Brake control system Rheostatic brake with electro-pneumatic brake system
- Bogie Bolster type with end beams. Solid axles.

Series 300 Introduced in 1992, this is the third generation Shinkansen used on the Tokaido route and has the following features:

- Maximum speed 270 km/h
- Passenger capacity 1323
- Weight/train set 710 ton
- Body material Aluminium alloy
- Vehicles/set 10 motored, 6 trailer
- Traction motor AC 3-phase cage asynchronous motor, 40 motors/trains
- Traction control VVVF (Variable voltage, variable frequency control)
- Number of pantographs 2/train set.
- Brake control system Regenerative braking, with electro-pneumatic brake system.
- Bogie Bolsterless without end beams. Hollow axle wheelsets

Between the three generations of trainset the real evolutionary change in technology has occurred with the series 300. Primarily the benefits of aluminium alloy construction leading apparently to a 20% weight reduction compared to the series 100. This, coupled with AC traction motors and variable voltage variable frequency (VVVF) control, allow the train to travel at speeds up to 270 km/h. This demonstrates the steady and successful progress of technical development backed by proven service experience and extensive use of prototyping. In the case of the series 300 a prototype train accumulated nearly 400,000 km of prototype running providing valuable data which was fed into many features of the production trains before full revenue earning service of the 300 series began.

It is perhaps testament to the technical innovation that all three generations of this train are in active reliable service today, with the service diagramming reflecting the train pedigree. At peak times the trains depart Tokyo in an hourly (11) 1-7-3 service pattern with:

1 Nozomi (series 300) that stops at major stations on the route*
7 Hikari (series 100) that stops at selected stations on the route
3 Kodama (series 0) that stops at all stations on the route.

*Whilst the series 0 and 100 are no longer manufactured ,the series 300 is and accordingly this series may operate on Hikari or in very rare circumstances Kodama services.

Technical innovation has not been overlooked on the conventional lines of Japan Central either. Conventional lines are a 1067 mm narrow gauge rail system (the original gauge of JR Central) which has seen the introduction of tilt technology with the series 383 EMU Shinano employing a controlled tilting system and self steering bogie. This has enabled both faster journey times and improved passenger comfort to be obtained.

2.4 Degree Of Technical Risk

There are many methods of controlling technical risk. Perhaps the most proven method is by use of prototype vehicles.

Presently this is being demonstrated on Japan Central with the 300X project. The concept vision of the project is quite simple 'to achieve the development of the best and most advanced high speed rail system using wheel on rail technology'. Some key development goals of the project being to:

- Improve aerodynamic characteristics
- Provide greater ride comfort
- Introduce comprehensive anti-noise measures.

The series 300X is effectively being used as the test bed for the next generation Shinkansen. Some developments to date include :

- Anti-noise features - Smooth bodyshell, aerodynamic noses, pantograph noise reduction covers.
- Ride comfort - Vehicle damping between car bodies, body tilting, active vibration control.
- Braking - Lightweight and highly resistant composite disc brakes.

In recent testing the series 300X achieved a Japanese speed record for rail of 443 km/hr. The previous Japanese record of 425 km/h was attained by the Star 21 of JR East.

The benefits of the series 300X will be incorporated in both the series 500 and 700 Shinkansen. These trains are being developed in co-operation with JR West, and will be the successor to the series 300.

JR Central however recognise that the Shinkansen is now being pushed to its technological limits, with operation at speeds above 300 km/h possible but ineffective in terms of cost performance and environmental damage such as noise and vibration.

Their solution to this problem is to examine the feasibility of constructing a second line between Tokyo and Osaka, this time with a Super Conducting Maglev.

The Super Conducting Magnetically levitated motor may be a radical solution, and could by comparison be as radical as the decision taken to build the Shinkansen in 1955.

It really is a case of looking back to the past, for the concept of magnetic levitation which was originally developed as a possible transportation concept in the early 1960's.

The Chuo Shinkansen development has already witnessed the achievement of a manned high-speed record of 531 km/h on the Yamanashi Maglev Test Line, on 12th December, and an unmanned record of 550 km/h on 24 th December.

Figure 4 - Maglev Prototype on the Yamanashi Test Track [1]

Super conducting Maglev technology offers a variety of advantages including the ability to withstand earthquakes (because of where the new line will be routed), such that the line remains safe for operation at high speeds.

A target of April 1999 through to March 2000 has been established for confirming the practical application of Maglev technology to regular high-speed operation.

From the viewpoint of transportation professionals, the development of the Maglev represents a radical step forward in both Railway technology and planned railway operation, and could potentially deliver the benefits of this transportation mode which have been promised for so long.

Already the test programme has witnessed developments in:

a) Super conducting magnet technology, leading to effective super conducting coils.

b) Car body structural design resulting in aerodynamically efficient, structurally tough and lightweight body shells.

c) Disc brakes from carbon fibre and carbon resin. Aerodynamic brakes.

d) Truck (bogie design) allowing a comfortable ride standard to be achieved at the same time as transmitting the propulsion and levitation forces generated by the magnets.

All of these developments with the super conducting Maglev serve to illustrate the facts that whilst the step change may be radical, that the change effect is tempered and mitigated by the use of prototype vehicles and sufficient research and development testing time. A common theme which permeates this railway resulting in technical success.

2.5 Technology Transfer

Up to the present time technology transfer on the Tokaido Shinkansen has occurred in a series of iterative steps in the development of the series 0, 100, 300 and now series 700 Shinkansen.

Whilst this transfer of technology has occurred in Japan, technology transfer has occurred from Japan to other parts of the world.

This is particularly evident for the Australian Tilt Train, due to enter service in 1998.

This train incorporates tilt system and traction control technology, utilised, proved and tested on the Series 8000 EMUs on the conventional line system of the JR Shikoku.

Looking further into the future, the developments of Maglev technology on the Chuo Shinkansen may provide a cogent reason for developing this in other railways around the world in the next millennium.

3. QUEENSLAND RAIL TILT TRAIN

3.1 Background

'**The most challenging Rolling Stock project in Australia** during this decade' is how the new tilt train being manufactured by Walkers for introduction into service on Queensland Rail in 1998 has been described.

This new train, which will be introduced into revenue earning service in 1998, is referenced in this paper because of the nature of:

a) The investment case.

b) Degree of controlling technical risk.

c) Technology transfer from railways in Japan.

Queensland Rail, which will own and operate the new tilt trains, is a State run Government Corporation and represents Australia's largest rail network. The passenger group comprises two sectors namely City Train which is an operation catering for 145,000 passenger journeys per day utilising fast EMU services around Brisbane and the Travel Train group, which operates long distance passenger services throughout Queensland. A very recent re-organisation has combined these two groups and it will be this organisation which will operate the new tilt trains. [2]

3.2 The Case For Investment

The Brisbane to Rockhampton route is one of the premier passenger routes in Australia. Serving customers to the Great Barrier Reef the railway is a narrow gauge construction (1067 mm similar to conventional lines on Japan Central and the gauge over which all tilting trains in Japan operate). Over the past decade this line has seen significant investment in infrastructure improvements, particularly electrification. This investment was particularly significant, because it was made during a period of economic downturn. Two benefits of this anti-cyclical approach were:

i) Large infrastructure and rolling stock investment projects implemented at competitive rates.

ii) Stimulation of economic recovery.

Following electrification the journey time for the 622 km route fell by 4 hours 30 minutes to 9 hours 30 minutes [3]. Almost immediately the service became more attractive to the customer because of the journey time reduction with the net result that over a 5 year period following electrification, passenger volume increased by 100%, co-incident with the new InterCity Express (ICE) trains manufactured by Walkers, introduced in 1988.

The route between Brisbane and Rockhampton is curvaceous and lends itself to tilt train technology. Sensing that further improvements could be made in journey time, with the resultant increased attractiveness of this service to the customer a case was made for investment in tilting trains. The investment case being supported by a study which predicted that the introduction of the tilt train would reduce the journey time from 9 hours 30 minutes to just under 7 hours, effectively halving the journey time since the infrastructure improvements began.

October 1994 therefore witnessed Queensland Rail awarding a multi-million dollar contract to EDI (Evans Deacon Industries), Hitachi and Itochu to construct two six-car tilting trains. The decision was taken to maintain the growth in passenger volume as well as supporting the economy, in this case tourism.

3.3 Degree Of Technical Innovation

For Australia the introduction of these trains will herald a complete step change in terms of the technology utilised. The trains will be formed of two half sets which will be arranged as a Motor-Trailer-Motor-Motor-Trailer-Motor formation and will incorporate many system features which will be unique for the operation in Queensland including:

- Active Hitachi tilt system capable of tilt up to a maximum of 5 degrees thus allowing a speed increase around some curves by up to 45%.
- Traction and Auxiliary systems featuring IGBT technology feeding 190 kW three-phase traction motors fitted to each of the four axles on each motor vehicle
- Microprocessor traction and braking control
- Pantograph control system (tilt correction)
- Aircraft style interiors including enclosed overhead lockers and entertainment systems.
- Fault management systems which monitor key vehicle systems.

- Lightweight construction and low centre of gravity.
- 600V DC Bus-Link for continuous auxiliary supply through neutral sections.

Because there are only two trains being manufactured, reliability and ability to 'turn the train around' aircraft style in minimum downtime are of vital importance to the operation for Queensland Rail. The degree of technical innovation must therefore be capable of allowing high levels of reliability to be established. Hence the installation of the fault management system coupled with redundancy of key systems built in to the train design.

Figure 5 - Tilt Train Vehicle For Queensland Rail

3.4 Degree Of Technical Risk.

The service introduction of these two tilting trains will merit detailed scrutiny within Australia. It is therefore vitally important that the design of the train minimises the inherent level of technical risk.

In this respect many actions have been taken to mitigate risk. These include the use of systems and technology which have a proven pedigree. Both the tilt system, bogie and Traction converters have proven a pedigree in operation on many networks in Japan. Furthermore failure in service is mitigated by the use of equipment redundancy.

Queensland Rail undertook a major reliability analysis utilising many techniques including Mean Distance Between Failure (MDBF), Failure Mode Effects and Criticality Analysis (FMECA). The conclusions of the study influenced both the selection of system components as well as determining the level of redundancy within the train. The level of redundancy being as follows:

a) Two separate refrigeration systems for each air conditioning unit, particularly necessary in a sub-tropical climate.

b) If the active tilt system fails the vehicle will revert to passive tilt.

c) Two auxiliary converters are designed such that if one fails the other converter will still comprehensively supply the services for the train.

d) Inverter redundancy for transformer pump and cooler motor.

e) Electro-pneumatic and regenerative braking systems are set-up to allow continual electro -pneumatic braking, even if the electro-pneumatic brake on one vehicle has failed.

The use of redundant systems in the key areas of the train offers striking similarities with the management of technical risk in the aircraft industry.

Further reduction of the technical risk will be achieved by the use of extensive test running over a 12 month period where the problems in the vehicle performance can be ironed out Effectively the first of the production trains will be a running prototype prior to its full introduction into service.

3.5 Degree Of Technology Transfer

Technology transfer has played a significant part in the design of the tilt trains for Queensland Rail, with Walkers the train builder, and Hitachi the system suppliers working very closely together.

In particular the bogie tilt system and the traction converter packages have all been utilised in operation on Japanese railways.
The tilt system utilised has been used on the JR Shikoku Series 8000 EMU, together with other DMU and EMU vehicles. In total some 547 vehicles since 1973 have been fitted with tilt systems [3].

By means of successful development and service experience this design of system has continually evolved from a passive to active tilt system. The active system being powered by means of servo-controlled pneumatic cylinders which are mounted between the body and the bogie, the cylinders working by means of a combination of intelligent commands from the control unit in each cab of the train. The control units are loaded with track data for the complete route and by monitoring train speed and location data from the AWS (station protection) magnets the leading unit instructs the tilt controllers on each vehicle. These controllers are responsible for servo valve control of the air supply to the tilt cylinders on each bogie.

The evolution of this system is a clear example of how a basic design with a high level of engineering integrity can be enhanced and continually developed.

4. ROMANIAN STATE RAILWAYS (SNCFR)

4.1 Background

SNCFR presents a diverse comparison between the other two examples of rail systems described in this paper, diverse because of the nature of the economy and degree of investment in Traction and Rolling stock.

With a population of 22.81 million and a land mass only slightly smaller than Great Britain, Romania is geographically well positioned at the cross-roads of Europe favouring North-South and East-West communication. Thus the development of a reliable railway network is viewed as being essential for a favourable long term economic outlook.

SNCFR operates freight and passenger services on a network which is closely integrated with other Railways. It has a total of 11,380 km line length of which 42% is electrified.[4] With the demise of communism and the advent of a capitalist market led economy, the rail system is embracing change wholeheartedly. This has allowed SNCFR to develop a corporate structure which facilitates the responsibilities of a public service obligation concurrent with targeting and achieving commercial objectives.

In achieving their commercial objectives the need to improve and invest in Traction and Rolling Stock was identified by SNCFR as a key factor.

4.2 The Case For Investment

The corporate modernisation of SNCFR is closely aligned with the modernisation of the company assets.

The key issue being the need to focus the modernisation upon improving operating efficiency. Thus the following items were identified as necessitating modernisation: [4]

- Infrastructure maintenance by means of implementing new technology in track maintenance.
- Upgrading of lines to allow operation at higher speeds (160-200 km/h) to allow integration into the European Transport system.

- Rolling stock modernisation with introduction of new improved vehicle interiors coupled with higher operating speed, features of which are expected to increase passenger traffic volume.
- Modernisation of Diesel and Electric locomotives to improve reliability, reduce maintenance and operating costs and increase availability.

The modernisation of the Traction and Rolling Stock is therefore viewed as a key element in the success of the restructuring plan. To maximise the available finance to support the modernisation and given the large numbers of vehicles in the fleet portfolio of SNCFR, the decision was taken to go for fleet refurbishment as opposed to purchase of new vehicles.

This presents an interesting challenge to manufacturers and operators, particularly in so far as refurbishment is concerned, where the introduction of new systems into old structures has the potential to not always give the degree of reliability desired.

4.3 Degree Of Technical Innovation

For the purposes of this paper three vehicle types pertinent to the refurbishment programme have been considered.

a) InterCity Coaches
b) 5100 HP Electric locomotives
c) Diesel Electric Locomotives

InterCity Coaches

The modernisation/refurbishment of the passenger coach fleet is of primary importance to developing business and maintaining an attractive standard of service.

100 coaches are due to be refurbished in 1998/99, with further large quantities planned in due course.

A key aspect of the refurbishment will be to increase the operating speed from 140 km/h to 160 km/h. This will mean installation of a new braking system as well as modification to the bogie. From a passenger perspective the interiors of the refurbished vehicle will be completely removed and replaced with:

- Air conditioning (1st class)
- Improved design of windows
- New improved seating design
- New interior fabrics and moquettes
- New toilets

Whilst the degree of technical innovation may be small, when compared to the Shinkansen, the benefits may be similar i.e. increased passenger loading.

Electric Locomotives

Figure 6 - 5100 KW Electric Locomotive [4]

To support the operation at 160 km/h, twenty-four 5100 kW Electric locomotives are to be refurbished with:

- The installation of a three-phase static converter for supply of auxiliary services
- The replacement of main switch gear
- The installation of a high-speed pantograph (160 km/h)
- The installation of contemporary WSP equipment
- Installation of a fault diagnosis system

Diesel Electric Locomotives

Seven Diesel Electric locomotives are to be significantly refurbished with:

- Replacement of the present diesel engine with a fuel efficient and reliable maintenance friendly unit
- The replacement of the main generator with an AC alternator
- The installation of an electric train heat supply
- Provision of an on-board health monitoring system

All three types of refurbishment will be based upon performance specifications which critically identify reliability availability and maintenance improvement criteria which the equipment will be expected to satisfy.

Significantly to support the modernisation and refurbishment five depots are to be modernised with installation of wheel lathes, test, and diagnostic equipment.

4.4 Degree Of Technical Risk

Technical risk is controlled by several means for the refurbishment programme.

For the coach refurbishment the principal means is by virtue of refurbishment of a prototype vehicle. This serves many purposes including:

a) Demonstrating the achievability of the specification and feeding vehicle performance issues back into a production vehicle;

b) Demonstrating the aspirations of the refurbishment to the tenderers;

c) Raising the public awareness of the refurbishment programme.

Figure 7 - Prototype Coach [4]

Indeed this method of demonstrating refurbishment has been used in the past in the United Kingdom, for example the HST Market Testing Vehicles (1993), although the full refurbishment programme in this instance did not progress into full scale production because of the privatisation of British Rail.

Similarly, for both locomotive types, various systems have been trialled prior to the recommendation for implementation/inclusion in the refurbishment specification.

The risks of failure are controlled by the aforementioned measures.

4.5 Technology Transfer

Technology transfer into Romania, has been very real for the refurbishment programme. In some senses this is inevitable given the multi-national nature of the Companies who are involved in this work.

There is however still the challenge to the manufacturer to make it work, with the uniqueness of each railway operation around the world leading to its own issues. What may work on a piece of equipment on a locomotive in say one country may not necessarily work in Romania.

The key areas where technology transfer is occurring is in the systems of:

a) Static Converters

b) Fault diagnosis systems

c) Diesel engine technology

d) Air conditioning technology

5. CONCLUSIONS

From the three railway systems studied from around the world, many common factors have emerged, particularly in the context of the new trains, which should be borne in mind for investment by owners/operators/manufacturers in new railways within the United Kingdom. The following conclusions can therefore be drawn:

- Investment in new trains is a growth catalyst for industry and economies not just in the local Rail industry sphere, but also in the wider picture relative to Country/State economies. For the United Kingdom the economic/social benefits may be seen initially in the local economies of where the trains are manufactured but the real benefits will occur when these trains enter revenue earning service as the example in Japan has shown.

- The degree of technical innovation follows a steady path with particular emphasis on evolution as opposed to revolution. This is supported by extensive use of prototype vehicles an aspect of which in the United Kingdom could prove beneficial if utilised. The onus should be on the Manufacturers to develop prototype vehicles, (similar to the aircraft industry), such that the quality/performance of their product is not compromised in service. Failure to utilise prototype vehicles may have an adverse effect on the ability to achieve full performance upon introduction into service.

- If revolution is required then the technical risks at the outset may be high, but with clear professional direction mitigation can be achieved by virtue of Research and Development focused on clear objectives.

- Finally as the railway world appears to expand, major manufacturing companies become larger by merger, so the advent of a number of common train types for the world becomes a reality. This could be further accentuated by procurement strategies being developed world-wide which are based on proven systems with a high degree of integrity specified on the basis of performance life cycle cost parameters, as opposed to one-off bespoke railway vehicle specifications.

6. REFERENCES

[1] Central Japan Railway Company Data Book (1997)

[2] Australasian Railway Association Yearbook

[3] Railway Gazette International - Jerry Jirasek/Don Heumiller

[4] SNCFR - Romanian State Railways (1996)

C547/010/98

The business case for new trains – developing a robust case

R CLUTTON MBA
Michael Hamlyn Associates Limited, Reading, UK

Despite pressures for the introduction of new fleets, new rolling stock will not be procured if it cannot be justified economically. However, the underlying case for new rolling stock may be better than might be expected if using a too simplistic approach.

A robust method for assessing the economic viability of new rolling stock is required which captures the myriad of variables, financing options, tax and pricing policies as well as the less tangible commercial considerations. Uncertainties relating to estimating accuracy, production risks and whole life costs need also to be assessed to ensure that analysis and conclusions are sound.

1. INTRODUCTION

Irrespective of whether or not the above thesis is accepted it can reasonably be argued that justification for new rolling stock is not as simple as it may first appear. Any business case assessment method and supporting economic model needs to be sufficiently complex to describe accurately the environments in which new rolling stock will be procured and operated - paradoxically, it must be simplified to a level which is usable.

This paper describes an innovative approach to the development of a business case model which takes account of the corporate and engineering processes involved in the provision of new rolling stock. The aim is to selectively introduce key issues and build up an approach to their integration in a business case model.

In practice, the issues, and particularly the risks, will be dispersed along the supply chain and the assignment of risks will be dependent on the specific forms of contract and terms applied. Here, the method and model are generic and would need to be enhanced to incorporate any specific characteristics in a practical case.

As the approach given in this paper is generic, and not least for commercial reasons, it has not been the intention to raise any specific issues relating to current or recent contracts. Any apparent references are thus purely coincidental.

2. STRUCTURE

The paper is set out as shown in Figure 1 below.

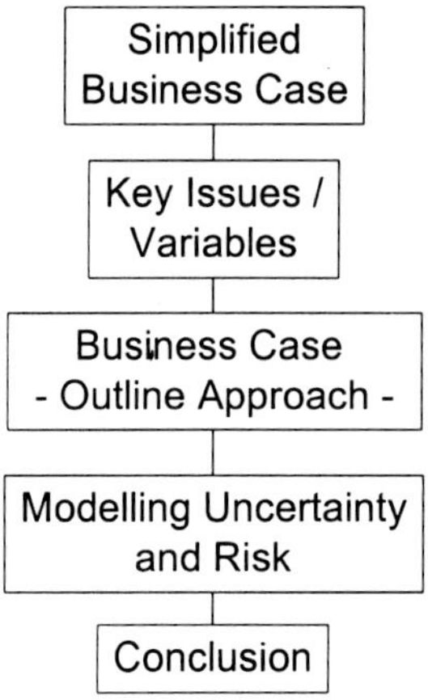

Figure 1 - Structure of paper

3. THE SIMPLIFIED BUSINESS CASE

At the time of privatisation of the ROSCOs there was much discussion regarding the viability of procuring new rolling stock. Due to the relative uncertainty in most aspects of the market place at that time, the economic case was at best awkward to define. The known variables were existing income streams from non capital and capital rentals and the "logic" implied by the Indifference Pricing Model. The best estimates of the market price for new rolling stock could also be argued to be a relatively known variable.

The simplified business case compares the present value of existing capital lease revenue for existing stock against the likely cost of its replacement. Not surprisingly the case is unsatisfactory to say the least. This simplified case, which we must recognise would never have been used in practice, provides an overtly pessimistic view of the potential for new rolling stock projects. We can be certain, because new rolling stock is being procured, that the case has to be better than this; or can we?

4. KEY ISSUES IN MAKING THE CASE VIABLE

4.1 Variables excluded from the simplified case

The various simplified business case arguments made circa 1995/96 were never presented as being robust; indeed, making a comprehensive argument was not in the best interests of all in the industry. All examples were caveated with known missing variables, such as:

- improved capacity;
- better reliability;

- lower heavy maintenance costs;
- discount rate sensitivity.

Also missing were:

- the capital allowances which can be derived from the investment;
- reduction in running maintenance costs;
- operational savings;
- improved safety performance;
- track access savings;
- passenger volume benefits;
- subsidy for new rolling stock (penalty for failure to replace old stock);
- the corporate (intangible) benefits, of 'flagship' deals for example.

Some of these items are hard to quantify and some may represent a dis-benefit rather than a benefit; nor do they accrue to one party in the supply chain. Nevertheless, they do substantially alter the business case and are required to bridge the financial differential between the existing and proposed leases. I suspect operators do not necessarily agree, but where the operator reduces his costs, and improves capacity or passenger appeal to a level where increased revenues are generated, the case for higher rentals can potentially be made. Similarly, if capital allowances are calculated and added to the equation along with any other subsidies then the viability of a case is again improved.

4.2 Indifference pricing

I remain uncomfortable with the Indifference Pricing Model. Whilst the model satisfied a need in the privatisation and valuation process and is conceptually robust, its application is, in my opinion, too arbitrary. We know that the capital value of new rolling stock will be greater than that of existing old stock, we can also be reasonably confident that the cost of maintenance will be improved. However, I do not believe, that the changed mix for new rolling stock is financially neutral when compared with the old. Moreover, in the simplified case, where capital rentals were used as a benchmark to the viability of a new rolling stock option, we would need to recognise specific weaknesses in the capital valuation.

For this reason we cannot readily rely on the capital income stream for existing leases as a valuation tool. If we need a benchmark for the economic argument we may be better placed if we make appropriate adjustments to the 'indifference' figures or look elsewhere, e.g. comparison with an optimum though hypothetical train.

5. THE GENERIC BUSINESS CASE - OUTLINE

Firstly I shall need to provide a definition of risk and sensitivity in the context of this paper. Broadly they are the same, insofar as they characterise uncertainty. However, the use of risk is more usually (though not necessarily) used where we expect a downside effect from our uncertainty; in a quantified sense, Probability x Consequence. To confuse matters slightly, a risk based model is designed to deal with both downside and hopefully upside effects of uncertainty.

5.1 Provision of accurate estimates

We can all have a fair idea of the likely range of estimating sensitivities for any given scope of work or specification. The larger the scope of work, the more variables we are attempting to contain within estimating limits. Arguably, the larger the portfolio of uncertainties, the more reliable is the total estimate; swings and roundabouts and so on. However, there is a propensity to over-estimate cost; the converse can also be true of course. The only way to gain protection from the portfolio is to have a more structured approach to estimating which allows the range of likely values for each estimate to be described. For example, this could be in the form of a distribution of cost against probability, instead of a static value which restricts access to the estimator's knowledge. On balance we can then be quite confident in the resulting figures. Indeed, we can determine statistical confidence in the results.

The potential volatility of specification, commercial costs and time-scales associated with new build projects mean that there is an expectation for estimates to be substantially different from reality. At the risk of sounding pessimistic, we can all site examples of profound difficulty in managing engineering change, holding time-scales (particularly the period from contract award to first train delivery) and the occurrence of unplanned commercial costs. In my experience this 'certainty of finding estimating discrepancy' traditionally results in arbitrary contingency being applied. This is a second level contingency in addition to that which may have been added by the estimator. This results either in un-quantified risks being taken or unnecessary cost being built into the project. It is also my experience that whilst an estimating engineer has to settle for specific values for his estimates, he invariably knows in detail the characteristics of the sensitivities and risks he is trying to balance. The same is true for estimators of commercial and overhead cost that are being incorporated into the business case. Unfortunately much of this valuable knowledge remains in the estimator's head.

If we can transfer, in summary form, the understanding of the detailed risks and sensitivities with which the organisation is going to be grappling, then senior management is better placed to strategically and tactically manage these uncertainties. By using risk analysis software superimposed over the business case model, we can capture the estimator's understanding of the dynamics of his estimates, and convey this to other contributors to the case and ultimately to the organisation's decision makers.

I do not intend to present the risk assessment process in detail as there are well-documented techniques for undertaking this type of analysis. In my experience this type of approach provides greatly improved estimating accuracy and helps the estimators make unbiased cost decisions. It also enables a case to be developed which more reliably reflects the required risk profile of the contracting organisation.

This risk based approach can be incorporated into the business case development process and model; it can deal with all relevant sensitivities that are discussed here. The capability of information technology, supporting a risk based model, enables us to understand for example, the impact on capital allowances and changed income profile of a slippage in (or acceleration of) the new build programme.

5.2 Elements of the generic business case

This section describes some of the key variables which we need to consider in the robust business case. It might be argued that certain variables are well understood, well documented and are already robust. I might generally agree but, for example, the apparent arbitrary use of a 'standard' discount rate widely exists.

I prefer to look at the business case as four sub models:

1 Lease Model - looks at the lease income over the life of the vehicles. Methods of calculating the initial lease price of new stock should be addressed using either a top down or bottom up approach and are briefly described later.
2 Whole Life Costing Model - which deals with the life cycle costs of the rolling stock asset running maintenance and Heavy maintenance costs of new rolling stock over vehicle life. This should take a balanced view of maintenance risks, i.e. risks need to be quantified, not as worst case, but as part of a portfolio of uncertainty. A risk based model facilitates this.
3 Capital Model - considers the cost of financing the new rolling stock over the vehicle life including the timing of payments and capital allowances. The effect of production run timings should be considered.
4 Profit Model - combines the three models above to assess the overall feasibility of the option being modelled. This model should allow for differing views for depreciation, residual value and residual risk. The performance measures for the business case, whilst similar in most cases, will usually be dictated by corporate benchmarks; payback, ROI, NPV/IRR, etc.

Various elements feed into and drive the four sub models. The interfaces between the elements and the four sub models are shown below in Figure 2:

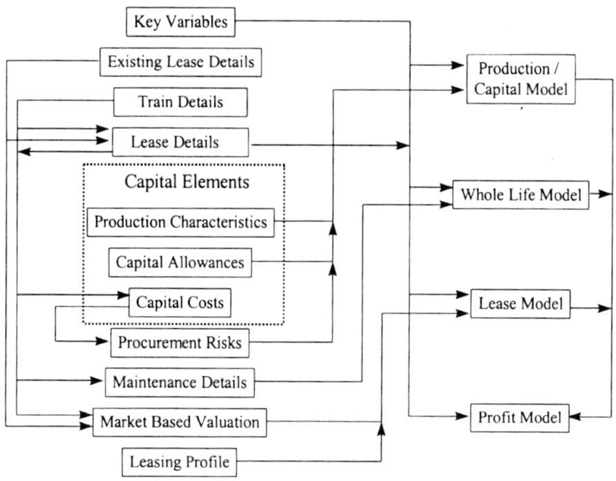

Figure 2 - Model interfaces

5.2.1 Key variables

Key variables, as defined here equate to those high level values, often set corporately, which are used widely throughout all elements of the business case model.

As I have just mentioned the arbitrary use of discount rate is surprisingly common practice. Whilst some simple discount rate sensitivity tests are undertaken in support of a case, the figures employed do not always reflect the risk characteristics of a given project or how averse the concerned organisation is to the project's risks. More often than not discount rates are the corporate standard and used for all projects.

We also need to include the risk free cost of capital to calculate the financing charges in the deal. The sensitivity of this figure in the broader economy will be determined to a large extent by the way finance is raised. Integrated with this or considered independently, we may also need to build in an exchange rate risk if the project is being financed in sterling but procured in another currency, the Euro for example. It is to be hoped that the financial expertise now so close to, or embedded in, the industry will provide a vehicle for hedging this risk. However, someone's model will have to include it.

Consideration must also be given to the effects of alternative depreciation profiles, e.g. straight line, accelerated. Other key variables are shown in Table 1.

Table 1 - Typical key variables

Discount rate
Cost of finance
Target profit margin
Depreciation term (years)
Market based or cost plus lease pricing for capital investment model
Straight line funded (or with balloon)
Alternative finance term (years)
Balloon value at end of finance term (% of capital cost)
Average supply side inflation
Average demand side inflation

5.2.2 Lease model

The lease model must endeavour to describe a likely pattern of future leasing; no mean feat. However this is possible to model probabilistically, and whilst the further into the future we look the more difficult the estimate becomes, the effect on the business case reduces. Adopting a risk based approach deals with the broad sensitivities that can be expected here.

The future leasing pattern can be described by the number and duration of future leases, the probability of securing these leases with the proposed rolling stock and the volatility of future lease prices. We must also take a view of likely mileage to feed into our life cycle costing model.

There are two possible methods for building up the initial lease price:

- Traditional (or Cost +) Method - where the capital and maintenance costs are determined, and used to derive a required lease price with a specified profit margin. This method can be used to derive a lease price 'cold', disregarding any external circumstances.
- Market Based Method - can be used to determine the lease price based on the lease price of the stock being replaced. This perceived market value of the rolling stock is based on:
    - the price the lessee is currently paying for rolling stock to provide a given service;
    - the various savings and benefits which can be attributed to the new rolling stock

A comparison of the two methods is shown in Figure 3.

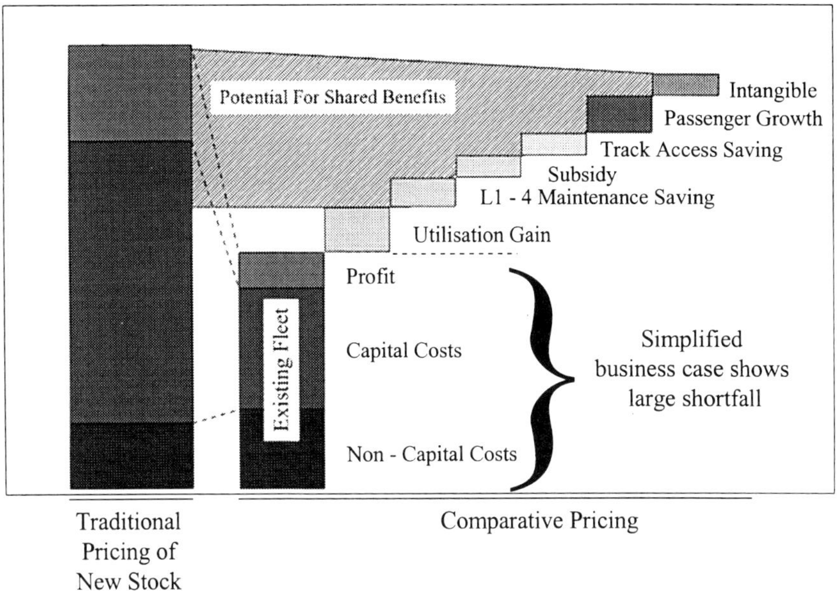

Figure 3 - Alternative pricing methods

As the new rolling stock is likely to be more expensive to lease than the existing stock the additional benefits are needed to map between the two to justify, as far as possible, the increase. There may be some projects which are regarded as less price sensitive, although it can then be argued that there is a large supplementary element needed to map between the two prices, e.g. relating to obsolescence of existing stock.

In many instances the benefits, such as a TOC subsidy (or dis-benefit of a penalty), will not be easily quantifiable, but again uncertainty can be modelled.

Typical variables required for lease modelling are proposed in Table 2. The distribution of these gains along the supply chain would have to be described depending on who was undertaking the analysis. It should be noted that the utilisation gain is the product of the reliability gain and the capacity gains of the new stock. It is from this value of utilisation gain that the number of new vehicles required to replace the old lease can be determined.

Table 2 - Typical lease model variables

Existing Lease Details
Number Vehicles in old lease
Trailers per unit
Powered vehicles per unit / Loco's
Vehicles per unit in current lease (calculated)
Current total lease income
Av. lease price per vehicle (calculated)

New Train Details
Number trailers per unit
Powered vehicles per unit
Number vehicles per unit (calculated)
Reliability gain
Capacity Gain

5.2.3 Whole life costing model - non capital considerations

This assesses the cost of Heavy maintenance for the new stock. The model employed must differ from the typical approaches I have seen because the model has to automatically assign the maintenance costs over the life of the vehicle when inputs are entered for each repair category. This is necessary for the risk analysis to be feasible. Typical inputs are shown in Table 3.

Table 3 - Typical whole life variables

Cost of repair
The number of repairs per trailer / motor
The basis of repair frequency (time or mileage based)
The repair frequency (years or miles)
The average mileage of the stock per year (linked to Lease Model)

The model used has to be self updating if we are using a dynamic risk model. For example, when the expected mileage run under the terms of the lease, or if periodicity for a repair is uncertain, the model adjusts its output when the input data is varied. This is fundamentally important as maintenance costs contribute so substantially to the case.

The relative advantages and disadvantages of this approach are summarised in Table 4.

Table 4 - Advantages and disadvantages of a dynamic whole life costing model

Advantages	Disadvantages
• data can be generated quickly and with less likelihood of introducing errors. • introducing iterative changes to the whole life model 'what ifs' are made more efficient, e.g., a change in average mileage introduces an immediate global change in the whole life model. • inclusion of risk based sensitivity analysis would be impossible in any other form.	• the alignment / misalignment of repairs has to be considered carefully at the point of data entry; e.g. alignment of bogie repairs with periodic interior overhauls. • where repairs have the same repair interval but are not aligned, there is no effect for new rolling stock but the approach is awkward for existing vehicles.

5.2.4 Capital model

The model must take into account the fleet introduction programme and the financing options. The analyst may need the option to allow for capital allowances as a benefit within the project, or to be excluded and dealt with at the corporate level.

The output from the capital model is used to assess residual risk in the profit model and forms an input to the traditional form of the lease price build up.

In order to describe the capital environment as accurately as possible it is necessary to include a view of the production characteristics as shown in Table 5. Of course, these can vary dramatically depending on specific contracted terms, the actual variables must be assigned on a case specific basis. Procurement risks also need to be assessed and incorporated.

Table 5 - Example production characteristics

Total manufacturing contract duration
Contract award to first train delivery time-scale
Contract award month
Average rate of production following train 1 delivery
% of capital cost to be paid on contract award
Milestone 1 at contract award plus X months
% of capital cost to be paid at milestone 1
Milestone 2 at contract award plus X months
% of capital cost to be paid at milestone 2
% of capital cost to be paid on first train delivery
% of capital cost to be paid on production deliveries
Retention to be paid X months after production end
% Retention

Additional data is required to describe the manufacturers learning curve, which is again a case specific detail. The contract award month is a driver to the timing of capital allowances. This is not required where a company with large capital expenditures establishes monthly companies to maximise the inflow of allowances. I am not aware that this exists, or is imminent, in our industry (yet). Also of course, there must be taxable corporate profits to benefit from capital allowances.

5.2.5 Profit model

This model requires little discussion, it is simply the residence for the results. The cash flow and performance measures are built up and calculated here. Residual risk and residual value are also determined.

The profit model must be able to capture the dynamic outputs from the other sub models, i.e. changes in capital expenditure, the income stream plus the changes in the maintenance programme and so on, when for example we have sensitivities applied to the manufacturing programme.

6. RISK MODELLING

In my opinion the financial analysis for this type of complex business case has to be produced using a risk based approach. A typical process uses simple spreadsheet technology but supported by secondary software for carrying out the risk (sensitivity) analysis. The software allows key uncertainties to be modelled dynamically with inputs based upon distributions rather than upon fixed estimates. By using probabilistic input data, the estimating uncertainty, as well as known areas of project financial risk, can be accounted for as far as is reasonably practical. This is carried out by varying the input data based on their defined distributions and sampling the output. This iterative process can be run for several thousand cycles, for example, or until there is statistical convergence of the output data. The output aids the capital investment appraisal process by defining the range of possible financial outcomes and the confidence of their achievement.

6.1 Input data

Probabilistic input data relating to cost estimates and quantities can be entered using distributions of various forms, e.g.

- Normal, a normal distribution defined by mean and standard deviation may be used but as estimates can rarely be described in this form more simple, but equally effective distributions are more appropriate.
- Triangular, where there is a spread of costs but a high probability of a specific value (not dissimilar to the mean in a normal distribution).
- Discrete, where the outcome can be one of two or more specific values.
- Uniform, where there is equal probability of an outcome anywhere between two specified limits.
- General, where the outcome can vary in value and probability of occurrence across a wide range; often where the outcome is driven by more complex technical and commercial issues.

6.2 Output

The output from the risk analysis is in the form of a distribution, often normal or skewed normal in form. Typically results values are reported at the mean, 5%/95%, and 20%/80% confidence limits of this distribution; i.e. at 5% confidence there is only a 5% probability of achieving this outcome or better, at 95% it is almost certain that the outcome will be at or below the value shown.

6.3 Interpretation of results

The values adopted by an organisation will reflect its view of risk, and the purpose for which it is using the figures. The mean figures are often used by an organisation for project budgeting but values at the 80% confidence limits are commonly used for capital investment appraisal. If a very low risk project is required or if it is an isolated project or of considerable value a pessimistic view may be taken, at the 90% or even 95% confidence level. Alternatively, if the project is considered amongst a portfolio of similar sized projects, a more optimistic view may be taken.

7. CONCLUSIONS

In early 1996 I asked a question in a similar forum to this one as to how dependent was the business case for new rolling stock on the exit strategies of the then ROSCO owners. As expected the answer given to this question was inconclusive. However, any purely numeric business case obviously needs to be tuned to the business objectives. Now that the industry has begun to mature and as the competitive forces in the rolling stock supply market are beginning to take shape, the business case will inevitably be tuned accordingly.

C547/026/98

New customers for new trains

R McDONALD
Thames Trains, Reading, UK

For many years, from the 1950s to the mid 1990s, British Rail worked hard to maintain passenger miles. Though we felt good about this, it did, in fact, represent a significant fall in market share. There were a number of factors at work, some cultural, some historical but the most real was competing for finite government investment with schools and hospitals. There are more votes in schools and hospitals.

Now we have the chance to invest in the industry and grow market share.

Since investing in new trains and using them imaginatively, my own company, Thames Trains has grown passenger revenue from £35m to £69m in four years.

Where might extra travellers come from?

1. Commuters.
2. Business travellers.
3. Leisure travellers.

We can grow by generating new or lengthened travel opportunities or gaining business from other modes. The trains, both in design terms and what operators can do with them is the most important part of driving this growth.

Let's listen to what some of our potential travellers have to say.

Firstly, Brian Businessman, aged 42.
"I've not been on a train for years.

I've got a company car which I take everywhere. All I need is a road atlas and my petrol card and I'm OK. It's sometimes a bit difficult to park, and the roads are getting a bit busy and the Chancellor is getting a bit funny about company cars, so I might think about catching a train one day.
The trains would have to be **reliable**. It is important to get to meetings on time. Predictability is the key word and that's what's getting difficult on the roads.

Frequency is important. I don't know what time meetings will finish.

I would like to be confident that the train will be there on time and to be able to get the simple **reassurance** that it is; maybe my secretary could check on the internet. I really do not want to be standing on a cold platform waiting for longer than I need to.

I want to be able to **use my mobile phone** and laptop when I want to, but **not be disturbed by others** using theirs. I certainly don't want screaming kids around me.

It is essential that the **train is clean, especially the toilets**. I only want to spend my money with companies that are well run and get the basics right.

It would be good to see some **innovation**. I'm used to British Airways slightly improving things month by month; they really think about what I want. The odd thing is that their planes are Boeings like everyone else's, yet I do feel that they cater for my needs. On **longer distance** flights their planes are better, with real innovation in seat design and **passenger facilities**. I've never yet seen anyone use the "jetphone" on domestic journeys though mobiles are banned. Looks like a gimmick.

I want to be able to find a **comfortable seat**, with **room to do some work or read a book**. It would be good if the train journey could be thinking or wind down time rather than wind up time like the M4 is becoming.

I want to be able to expect the same standards of **catering** whatever the length of train. My journey between Bristol and Bangor has the same importance as the journey to London. I expect something between a full meal and a sandwich.

I'm used to things **looking good.** The interior of my Peugeot is very smart; the first time I sat in it I thought "Wow!" Trains should be the same. On a train I want to look down on the car drivers on the motorway, knowing that they are looking up at me thinking "He made the right choice."

When I get to my destination I want to be **reassured** that my onward journey will work. Where is the taxi rank? Is the Underground OK? Is my connecting train on time? There must be some IT solutions."

Secondly, Laura Leisuretraveller.
"I never travel by train. I just bundle the kids into the car and off we go. The traffic has been getting worse recently, but it's still not bad if you avoid the rush hours. I know lots of rat runs anyhow, to avoid the worst jams.

I'd like to travel by train. The **kids** would love it, for the first half hour anyhow, but the trains are too awkward.

It's difficult to get the kids, a buggy and luggage on to the train. Some of the newer ones are easier with big wide doors near the seats, but the Intercity ones are difficult, fighting your way past the toilets and the luggage racks looking for a seat.

You are never quite sure whether to scramble into the coach and bagsy some seats, with luggage and children flying about all over the place or get your luggage organised first. You sometimes get into a coach and find that some bright spark has decided that this is for business use only and the family coach is four coaches down the train. It would have been better if **I'd known where to stand on the platform.**

I'm never quite sure if our fellow passengers appreciate my children as much as I do. We get some funny looks from business men. They (the kids, that is) crawl about into all sorts of corners. It is essential that the **train is clean**. They do seem to design in all sorts of dirt collecting nooks and crannies. **Clean and working toilets** are crucial. When Georgie, aged two and a half wants to go, he wants to go now.

Some cheap sandwiches, snacks and drinks would save a lot of bother packing picnics. We are not interested in complicated sandwiches just **value for money**.

The same applies to the whole journey really. There are a lot of good deals, though they are a bit complicated. I don't always want to book in advance. I sometimes want to take the kids out on a sunny day on impulse. Other privatised industries have brought the price of things down but the train fares keep rising. They really do need to **get more efficient**.
Some **play area** would be great. Even my local Building Society has one now. My kids could really enjoy travelling with other kids.

I'm really concerned that I would **not be able to get off the train**. Information in the coach about train stops would be really good. I'm not thick but I would like some **reassurance**. I might not know when to stand up and get the troops assembled. That arrangement where you use the outside door handle seems really odd.

I like to see **disabled people** looked after, and facilities being easy to use. Some days we might want to take bikes on the train, but I don't know how good they are for that. They do seem to be designed with the single luggage-free traveller in mind.

I don't use the trains when I'm travelling by myself because, frankly, I'm **scared of being mugged** on the local trains."

Thirdly, Kevin Coolyoof, aged 19.
"I don't travel on trains because none of my friends do. We used to have a laugh travelling in a group when I was a student. The railway was part of the establishment, up there with headmasters and politicians. We used to nick the odd light bulb to wind up the ticket inspectors.

I've got an old Beemer now. Much better for impressing my mates.

I do get a bit worried about the **environment** and all that stuff but you only live once, don't you?"

Finally, Carol Commuter, aged 36.
"I've travelled for five and a half years on this line, and paid £5,000 to British Rail or whatever they call themselves nowadays. There's no way I'd use the train in the evening or at weekends.

The trains are **dirty and impersonal**. It's the most sordid ten hours of my week.

I don't ask for much. The train should be **reliable**. I only really want something as effective as a horizontal lift, getting me from A to B. Ninety percent "Charter" performance means that I am seriously late once a fortnight.

Price reductions like BT and even British Gas have made since privatisation would be welcome. When I started commuting ten years ago I had a Cortina. My current Mondeo is a much better car, and taking inflation into account was cheaper

I want a **comfortable seat**, not too close to my fellow passengers.

Things will go wrong occasionally. I then need to be **kept in touch** with what's going on.

I feel like an exploited victim not a customer getting value for money. I do have choices, I am thinking of moving house or changing jobs to get away from this."

Conclusion.
Is this too far fetched? We have 6% market share. It probably represents the views of many of the 94% who don't use us and many who do.

As an industry, whether operators, Railtrack or train builders and maintainers we must develop a vision and a confidence in being a growing industry by giving a predictable, clean, easy to use, value for money railway as a minimum.

We somehow need to make train travel aspirational.

Performance specifications – what are they for and what must they really contain?

P J RUSSENBERGER
Traction and Rolling Stock, Halcrow Transmark, UK

SYNOPSIS

After establishing the nature of Performance, this paper suggests that there are four fundamental types of Performance. Possible characteristics of and functions for Performance Specifications are proposed and from these ideal contents are discussed. It is proposed that these should be kept to an absolute minimum to allow the suppliers and their engineers freedom to provide Rolling Stock which will fulfil the purchaser's needs and provide value for money without constraining them within preconceived ideas. Finally, it is suggested that the Performance Specification is all that is needed in laying down the description of new Rolling Stock.

1. INTRODUCTION.

Specifications have been developed for at least as long as there have been engineers to build equipment to meet them. The earliest specifications may have been little more than an idea in the mind of the engineer, perhaps derived from a conversation with an entrepreneur seeking to make his fortune from the business of a coal mine, canal or railway. Today the production of new equipment is accompanied by the production of Business, Particular, Performance and Procurement Specifications, all of which form part of an Invitation to Tender. This paper considers the issue in relation to railway rolling stock.

2. WHY OBTAIN NEW ROLLING STOCK?

New Rolling Stock is costly. Investment in a new fleet is not to be undertaken lightly; great care is needed in establishing the real purpose behind the investment in order to ensure that the purpose is met and good value for money obtained.

The overall objective of a company is to meet the aspirations of the shareholders. If it is a private company, these aspirations are usually to remain in long term business while delivering the maximum profit. A tier of government is normally the shareholder in a nationally owned company, which may well have its objectives adjusted towards cost reduction rather than profit maximisation in the delivery of a given product, but the aim of keeping financially healthy remains.

New stock is traditionally introduced as part of a wider product improvement, possibly involving an electrification scheme, or because the current stock is deemed life expired.

Product improvement will be driven by the desire to increase, or at least maintain, revenue and market share on a given route. It need not be limited to improved ambience or company image, but may well include alterations to the timetable such as reduced journey times, adjusted stopping patterns or service frequencies. These will influence the commodity to be delivered by each element of which the Rolling Stock is but one.

The replacement of life expired equipment is often initiated by a potential for controlling costs which can show an adequate return on the investment demanded. It is rarely limited to a straight like for like replacement as the opportunity can be taken to introduce enhancements, including new technology, which deliver a product improvement. A project initiated to reduce maintenance costs can expand to encompass product improvement and cause the original objective to become secondary. This is not necessarily the result of erroneous project development if the evaluation has shown that a better financial performance can be derived from the provision of significantly different equipment. The requirement to replace otherwise serviceable Mark I Rolling Stock on safety grounds is akin to the replacement of life expired equipment and its implementation can be expected to yield other improvements.

The recent British franchising process has introduced another reason for obtaining Rolling Stock. In at least one case it has been a condition of the award of the Franchise. Here it becomes a necessity for remaining in business. If the financial calculations have been done correctly, it should also maximise the franchisee's profit but the driving feature is the requirement to introduce new trains.

Whether it is to meet product enhancement, cost reduction or franchise compliance, it is the overall business strategy which must direct the Rolling Stock requirement and the decision to acquire new vehicles. Within that strategy the Rolling Stock will have specific objectives related to other aspects of the business indicating how the equipment must perform. These objectives will influence the tenor of any Specifications issued with Invitations to Tender.

3. WHAT IS PERFORMANCE?

An examination of the need for and the content of Performance Specifications presupposes that the concept of "Performance" is universally understood and defined. This is not always the case. A "high performance" car is usually assumed to be one which can accelerate rapidly to a high top speed. That type of Performance may sometimes be the prime consideration, but in other circumstances it is the number of occupants the car can carry or its fuel consumption that will be of greater importance.

This does not define "Performance" but, in passing, two points already arise. Firstly the performance required by generically similar equipment can differ; secondly, given the usually changing commercial environment, it can change with time.

Dictionaries provide a useful start with such definitions of the verb "to perform" as "to fulfil," "to execute a function" or "to carry through to completion." These lead to the noun "performance" being defined as "manner or success in working," or "the accomplishment, carrying out, doing, of any action or work" (1).

Performance concerns what is done, rather than how it is done. The end result is the matter at issue. How this end result is reached is important, but it should not properly be the concern of any measure or specification of Performance.

4. WHO REQUIRES THE EQUIPMENT TO PERFORM?

It would be easy, and truthful, to respond to this question with "The purchaser, of course." The purchaser undeniably requires performance for that is one of the reasons for obtaining the equipment. Something has to be done and the equipment is required to do it.

Given the situation where the Railway Operator is the Rolling Stock owner, the development of a Performance Specification is relatively straightforward since only the objectives of one organisation have to be fulfilled. Those objectives include the profitable sale of transport to the travelling customer who purchases the product the Rolling Stock is used to provide.

In considering the purchase of Rolling Stock for a British passenger railway, a further tension becomes evident. The purchaser and the end user are not usually the same bodies and, while they have to work together, their final objectives are unlikely to be the same. The function to be fulfilled by one may not necessarily be fully compatible with that of the other.

The purchaser is usually a Rolling Stock Leasing Company (ROSCO) which remains in business by leasing its Rolling Stock to Operators. The end user is normally a Train Operating Company (TOC) whose business it is to run trains using leased Rolling Stock and train paths. If the needs of the TOC cannot be met, the whole process must become untenable because the TOC will cease to lease the Rolling Stock. Similarly, if the ROSCO cannot obtain a reasonable rental for its assets, it will ultimately become unable to trade.

Thus there are different types of Performance and different objectives for each.

5. TYPES OF PERFORMANCE.

While the exact boundaries between them may not necessarily be clear, the types of Performance a train can be required to meet can be divided into four aspects:

Safety
Technical
Operating
Financial

The advantage of making this particular division is that it enables the Purchaser to examine the individual aspects of Performance in relation to his whole business strategy. It is worth considering each of these aspects in isolation to question and establish what it really needs to contain.

5.1 Safety Performance.

Arguably, the most important aspect is the Safety specification. The behaviour of Rolling Stock in the event of a mishap such as a derailment, collision or fire is vital to ensure the survival of as many of the occupants as possible, each sustaining the least injury. It is often limited to vehicle strength or energy content of combustible material, but should it also include at least an indication of the evacuation rates the layout of passenger saloons could achieve? This is, after all, a fundamental requirement for aircraft certification.

It is conventional to specify Safety Performance in accordance with the Group Standards of the Track Authority and the Operator. These usually assume that a mishap will take place and that the vehicle performs so as to contain the effect immediately after the mishap has occurred. There is an unconscious definition here of a mishap as, for example, a collision whose effect is damage to Rolling Stock. Should the mishap be defined, say, as a signal passed at danger whose effect is a collision? This would leave the engineer free to provide a system such as track brakes which could provide an emergency braking rate in excess of that delivered through the wheels in order to mitigate the effect of the mishap.

A simple statement of compliance is rarely sufficient to convince a Track Authority that a train can be used safely over its tracks. Proof is required that the train really does comply with the given Standards. Acceptance of parameters such as axle loading and swept kinematic envelope are simply proven by simple tests or during design scrutiny; freedom from interference with signalling and telecommunication circuits is more difficult to establish. A manufacturer may wish the Track Authority to specify limits which must not be exceeded content in the knowledge that keeping within these limits will allow Safety Certification to be achieved rapidly. Unfortunately, the Track Authority may require that proving tests take place to ensure that new trains meet the conditions, foreseen or not, to be encountered on the routes over which they are to be operated. There is a tremendous advantage for a purchaser in placing the onus for gaining acceptance on the supplier since this is a high risk area. By accepting the risk, the supplier is given a massive incentive to ensure his train is acceptable to the Track Authority. In the longer term this approach could lead to a Track Authority being forced to issue, and comply with, definite parameters to be met.

5.2 Technical Performance.
This is generally regarded as the engineering description. Technical Performance involves matters which, while they influence it, may be somewhat removed from the end product as seen by the purchaser's and user's customers. For example, the ability of a vehicle to survive a mishap has no direct effect on the passenger until a mishap actually occurs.

Availability and Reliability are linked. In an Operator's mind they (unfairly?) represent the shortcomings of the engineers since they concern situations in which the equipment cannot be used and may even be causing a costly disruption It is important that these parameters are specified in a quantifiable manner which can be monitored. The purchaser may have to decide whether he is seeking a number of diagrams to be covered or specifying a maximum permitted maintenance downtime within a given time period. In either event, Availability will be linked to the maintenance requirement and will provide the designer with the maintainability and reparability targets to be met.

This can be amplified to a complete maintenance strategy. The degree of detail given will depend on whether the purchaser is intending to be responsible for maintenance or to include it in the product being purchased. If it is the latter, then the supplier should be left with considerable freedom while being required to demonstrate that the proposed maintenance regime is efficient and part of a totally cost effective package. In either case, it must be a requirement to show that the specified Availability can be achieved. While it is likely that a balanced system of examination will be appropriate, but it may be unwise to make it an absolute requirement for compliance.

Reliability is often specified in terms of the mean time between failures of varying given degrees of hazard. These are usually described in terms of the effect on safety or the degree of danger to which the travellers are subjected. Although reliability is not now usually specified solely as distance run between delays in traffic due to an engineering shortcoming, merit still remains in targeting a limit on the effect of a relatively minor defect on the service being delivered. There is often a tendency to specify that "only proven equipment" should be used as a means of achieving a particular reliability performance. This is understandable; it is also unduly restrictive, not least because "proven" usually means "used successfully in squadron service for some years on somebody else's railway." The problem is that it hinders technical development through placing the entire risk and direct cost upon the supplier. Reliability prediction procedures exist to establish whether a new system is likely to achieve the required reliability.

Route Compatibility is fundamental to any Rolling Stock, new, refurbished or simply transferred. It is not only the classic kinematic envelope, but includes length (for platform compatibility), weight (both axle loading, weight per unit length and track forces), electro-magnetic compatibility with signalling and telecommunications systems and the ability to operate under traction or braking over any of the gradients to be encountered.

Often forgotten, but of increasing commercial rather than social and political importance, is the Environmental impact caused by the new Rolling Stock as the philosophy that "polluters must pay" gains credence. Apart from the usual prohibition on the use of harmful materials such as CFC gases, there may be a requirement to limit the total use of materials. This is easier to specify if a totally new railway is being built and, when such fundamental issues as the choice of traction system have to be considered as part of the whole railway system, it can extend to requiring the railway's visual impact to be minimised.

5.3 Operating Performance.

Theoretically this can be specified by the proposed timetable related to the route characteristics, such as maximum permitted speed, gradient profile or signal spacing, and the anticipated passenger flows and type of passenger amenities to be provided.

A prospective purchaser may well be very wary of such an approach. It appears too close to asking the builder what he would like to supply rather than indicating what is needed. In fact, it may not be.

The timetable and route will specify the traction package, performance and number of trains required in service. Correctly specified, the passenger throughput and amenity will define the seating layout, having due regard to such matters as the number of doors needed. This in itself will be partly driven by the timetable, since passenger boarding and alighting time will have its own effect on the station dwell time, thus affecting the overall running time.

To take another issue, passenger flows will relate to seat design. The longer the average journey time, the more comfortable the seating needs to be and the greater the number of seats to be provided in relation to the number of passengers carried. However, rather than specify a design of seat, the specification could merely require that "the 95th percentile of customers should be seated comfortably for [say] 45 minutes." Despite the definition of "comfortable" being open to debate, this leaves the supplier free to draw on his own experience and to develop his own design and layout.

Where there may be a need to lay down an apparently restrictive specification is in relation to other trains in operation on the route in question. This relates to operational technicalities and appearance.

Operational technicalities concern such issues as the ability to work in multiple with or enable recovery of trains of other types. This involves more than coupling compatibility. Achieving the correct performance in this area can prove critical to the overall performance of the railway, particularly in regard to providing the timetabled capacity in the event of disruption due to a cause external to the Rolling Stock, or containing disruption and speeding service restoration following a train failure. What is apparently a restricting requirement in the Performance Specification will provide the railway with greater operating flexibility.

The appearance of Rolling Stock is part of the overall ambience the railway presents to its clientele. While changes have to occur to enable progress and prevent commercial stagnation, the mixing of Rolling Stock of unnecessarily widely differing designs can give the impression of a chaotic organisation. For example, specifying external cab layout and body profile can significantly aid the presentation of a cohesive image, but designing a vehicle to suit its livery is to be avoided, since a functionally designed vehicle should be able to accept most soundly developed liveries. The Performance required here is that the Rolling Stock is to complement and enhance the image presented to the traveller by the railway.

5.4 Financial Performance.

All privatised railways are in business to make money. While part of the periodic assessment of the railway will include punctuality, service reliability and similar statistics, in the current British regime, shortfalls in these areas lead to financial penalties. The final measure of the company's performance remains its balance sheet. Neither the Technical nor the Operating Performance Specifications indicate in themselves how much Rolling Stock should cost to operate, yet the running expenses and interest paid on capital used to finance the purchase will inevitably contribute to the year end profit or loss of the TOC or ROSCO.

There is a temptation to specify first and operating costs which are not to be exceeded and to make these as low as possible. The very grave danger here is that this will constrain the final product to such an extent that it becomes unable to fulfil its Technical and Operating Specifications. The supplier will probably be forced to build down to a price rather than up to a standard.

In developing Financial Performance criteria, the place of the Rolling Stock in the whole profit and loss account has to be established. The effect of its introduction on ridership and farebox income has to be predicted, since in the extreme situation it is possible that this will rise so much that an increase in operating, maintenance, rental or capital costs will still produce a net gain for the company. The railway's financial objective has to be to maximise that gain. Here the optimum balance has to be struck between costs and revenue.

The ROSCO will also be seeking to maximise its financial gain. The costs encountered by a ROSCO are those of capital servicing, heavy maintenance and technical support. All these are regular outgoings. It is likely that a reduction in the need for heavy maintenance, and thus its cost, will be accompanied by an increase in the first, or capital, cost of the fleet. The sole source of income is vehicle rental. This can be increased per vehicle if the operating cost to the TOC is reduced or a lesser number of vehicles is required. Modern Rolling Stock requires heavy maintenance so infrequently that the benefit from lengthening the interval between repairs arises from cost reduction rather than an increase in rental income arising from increased availability.

Rental income has to be maintained over the whole life of Rolling Stock. Continually improving standards and changing commercial conditions mean that a fleet cannot necessarily be expected to remain on the same service or with the original TOC as when it entered traffic. The ROSCO therefore requires Stock to be re - leasable at the end of the initial lease and this may introduce a further conflict. The ROSCO will wish to make Stock as versatile as possible. This usually means giving it the widest possible route compatibility - for example short vehicles capable of working on both British electrical supply systems. It is unlikely that a TOC operating over routes capable of handling the longest vehicles and having only one system of electrification would be enthusiastic about paying a rental commensurate with such versatility. The solution is in the hands of the ROSCO, but it could include, say, specifying the ability to be economically modified to accept power from a different or additional electrical system.

5.5 Interactions.

It may not be appropriate in writing the Performance Specification to divide it strictly into the sections just outlined because they overlap. Any supplier will have to develop his own tender which will indicate his response. This should provide his optimum balance between the requirements of a fully compliant performance and lowest cost.

The contribution to the Financial Performance by the Technical and Operating Performances of any equipment can constrain its ultimate success or failure.

6. WHAT IS THE FUNCTION OF A PERFORMANCE SPECIFICATION?

The Performance Specification indicates what the Rolling Stock is required to do. It is developed by the purchaser and, if the Stock is to be leased, with at least a very significant contribution by the end user who may in fact have initiated the process.

In developing the Specification, both parties are forced to examine in detail what they seek from the Rolling Stock. One function, therefore, is to clarify the objectives within the relevant section of the business to be fulfilled. This is essentially internal to the purchasers. It is vital. Without it the purchaser cannot establish clearly how the Rolling Stock is to fit into his business strategy or, indeed, whether it fits. In its absence he runs the risk of acquiring unsuitable equipment or having his strategy unnecessarily constrained until some remedial action can be taken which will impose its own additional cost.

A second purpose therefore is to draw together the aspirations of the user and purchaser. Taking the British situation, there is no good reason why these should not initially be developed independently, though they will have to be reconciled before a manufacturer can be approached.

Thirdly, the Performance Specification indicates to the supplier how his product fits into the purchaser's schemes. This gives him the tenor of the equipment he is to supply, thus giving him a clear sense of direction in his design work.

Lastly, the Performance Specification defines the criteria against which any Tender will be evaluated. This ensures the emphasis placed by the supplier on any particular aspect is at least compatible, if not coincident, with that of the purchaser.

7. TENDER EVALUATION.

The Tender Evaluation stage of the procurement process serves to assess Tenders for their degree of compliance and to rank those which comply sufficiently to warrant further consideration.

There is a need to ensure equitable evaluation between Tenderers. This necessitates a marking system which is rigidly applied and has been developed before any Specifications are issued. Such a system should be weighted to favour those Tenders which have the greatest degree of compliance in the aspects most important to the purchaser. It is important that both this weighting and any specific issues over which a lack of compliance will lead to disqualification are disclosed to Tenderers so that they deliver responses directed towards the purchaser's preferences.

If carried out properly, this procedure will ensure that the place of the new train in the purchaser's overall business plan is understood clearly and also provides an opportunity to audit the part it plays. A further advantage is that the purchaser is forced to examine his requirements very carefully and to rank the absolute and relative importance of particular issues.

Part of the Tender will be the overall price of the trains to be delivered and the importance of cost can be indicated at this point of the Specification. To facilitate a cost/benefit study of differing designs for the same feature, a requirement can be given that the cost of certain sub-systems within the whole of the train must be described. The purchaser will be enabled to compare widely differing proposals. This matter assumes greater importance if a full maintenance contract is being purchased as well since an increase in first cost may be balanced by a reduction in the maintenance requirements yielding a higher Availability. If the fleet being obtained is large enough, this could actually lead to a smaller number of units being required, while the purchaser's needs are still met.

8. CAVEANT EMPTOR ET VENDOR – THE LEGAL ISSUES.

It has to be understood that any Specification will be taken to mean what it says; it must therefore say what it means. Clarity is vital. Any failure to supply what has been contractually agreed is likely to lead to a reaction quite possibly ending in litigation. Equally, in evaluating tenders, purchasers must be certain that they are making no assumptions about a Tenderer's response which are not clearly indicated. There can be no merit in questioning the failure to supply what has not been tendered.

9. WHAT MUST THE PERFORMANCE SPECIFICATION REALLY CONTAIN?

As far as the TOC is concerned, the contents of the Performance Specification should be clear from their business strategy. The Specification developed by the TOC should therefore be limited to the following:

- Safety -The compliance with Group Standards
- Availability - The diagrams to be covered and/or the proportion of the fleet to be available for traffic.
- Reliability - The failure rate described both by frequency of degree of hazard and by operational effect.
- Route Compatibility - Where the Stock is required to run and the Track Authority's specification for that route.
- Business Requirements - Passenger flows to be accommodated, timings to be achieved, passenger amenities required and overall image.
- Environmental Restraints - Any limitations of the environmental impact.
- Evaluation Philosophy – How the purchaser will assess and rank the Tenders received.

The ROSCO may wish to adjust some of these before the Specification reaches the manufacturer in order, for example, to facilitate subsequent re-leasability. This should cause no difficulties, but this part of the Performance requirements is clearly the ROSCO's responsibility and should have minimal impact on the TOC.

10. CONCLUSION.

All the features discussed in this paper relate to the end product. They indicate what the Rolling Stock is required to do; very little indication is provided of how it is to be done. This is correct because, given the definitions of Performance described earlier, anything else lies outside the scope of Performance.

If the Supplier is to be given the greatest freedom for the development of new approaches and innovation to meet the specified Performance, then the Performance Specification should be kept as brief as possible and be limited to those issues which are unavoidably required. It must be minimalist.

Yet the Performance Specification should state all that is needed to define the technical and operational purpose of the Rolling Stock. Therefore it is the critical one which is produced during the process of acquiring Rolling Stock and should be the complete indicator of what is being sought.

This begins to beg the further question "What else is needed to specify Rolling Stock?" A possible serious answer is "Nothing." If the Performance has been derived correctly, any other requirement is an unnecessary addition that constrains the supplier and may prevent him from providing his best product. The Performance Specification must impose the fewest possible constraints.

In terms of specifying the Rolling Stock, the Performance Specification may well be the only one which is actually required.

REFERENCES.

1. The Shorter Oxford English Dictionary and Chambers 20th Century Dictionary.

C547/025/98

New vehicles for continuing rail head conditioning and track treatment operations

M P GIBBARD BSc, CEng, MIMechE, ACGI
The Engineering Link Limited, Derby, UK

EXECUTIVE SUMMARY

This paper describes the work that Railtrack has undertaken to develop improved techniques for the cleaning of the running rails, the condition of the conductor third rail and also track bed and track side weed spraying. Railtrack has concluded that it needs some new vehicles and new techniques to improve the current performance particularly in the Autumn leaf fall season. The Author is Martin Gibbard, Engineering Contracts and Projects Director of The engineering **link** who have assisted Railtrack in various elements of this project to date.

1. BACKGROUND

1.1 Rail Head Contamination

For approximately eight weeks each Autumn, the train services that operate over the Railtrack infrastructure are vulnerable to leaf-fall. This occurs due to fallen leaves being swept up by the slipstream of passing trains and being caught in the nip between the train wheel and the rail.

The leaves are gradually rolled into a layer of crushed debris on the rail, forming an insulating coating. This coating can cause track circuit malfunctions and, when the layer is dampened by dew or light rain, becomes extremely slippery, reducing adhesion below the levels on which signal braking distances are calculated. Other problems that may occur include:

- Reduced traction performance due to wheel spin.
- Extensive wheel-flatting due to wheels locking when braking.
- Lengthened journey times due to cautious braking and wheel slip.
- Train cancellations due to shortage of stock whilst remedying wheel damage.

Much work has been carried out nationally by Railtrack in order to reduce the effects of the railhead contamination.

Treadles have been employed to supplement track circuits at level crossings, thereby ensuring that any failures to operate track circuits do not lead to accidents.

Site surveys have been undertaken on the national rail network which have identified particular locations that are high risk due to the types and numbers of trees near the line. As a preventative measure a planned programme of vegetation control is carried out which includes the cutting back of line-side vegetation and heavy pruning. This is designed to keep the track-side clear of the broad-leaved trees that cause the problems.

However, most of Railtrack's attention is concentrated on remedial action to reduce railhead contamination. The two most satisfactory schemes so far for countering the crushed leaf mulch are Sandite application and cleaning by high pressure water jetting.

Sandite Application
Sandite is an adhesion improver laid along the railhead from specially equipped vehicles. It consists of sand mixed with a gel which is applied to the rail. Some Railtrack zones use a variation called Sandite S4, which also contains stainless steel shot. This is added to improve the operation of the track circuits. Known problem areas are targeted, particularly where there are deep cuttings and steep inclines. However, the benefits from Sandite application are relatively short-lived as much of the gel is removed by passing trains and re-application is required daily or more frequently.

High pressure water-jetting
High pressure water-jetting is now being used to a very limited extent on two zones. This removes the leaf mulch by means of a controlled high pressure jet of water, before the leaves form a hard coating. Water jetting and Sanditing can be carried out at the same time from rail vehicles.

1.2 D.C Conductor Rail Icing

As soon as the Sandite season has finished and sometimes beforehand, another problem occurs which is ice forming on the conductor rail. This affects the third or fourth rail system which is in operation in parts of Railtrack Southern, North West, Midlands and East Anglia Zones.

Ice forms a barrier between the live rail and the collection shoe and causes arcing and can lead to, at worst, damaged traction equipment and fires. To prevent the ice forming, conductor railhead scraping is carried out followed by the application of a de-icing fluid. This fluid is a chemical which can be applied to the conductor rail to keep it free of further ice build up.

With the passage of trains, however, this is removed and again as for Sandite needs to be replenished at least daily.

1.3 Track-bed and bank-side weed growth

Other track treatment functions that are required during the Spring and Summer are weed-spraying activities which include herbicides being sprayed onto the track and bank-side areas to control the growth of weeds.

2 FEASIBILITY WORK

The above activities are currently undertaken by various vehicles and locomotives that have been specially modified and equipped for the work.

Many of the vehicles which are owned by Railtrack are former surplus passenger carrying vehicles and are up to forty years old and have come to the end of their useful life. The condition of the vehicle body-work is frequently poor and components are in need of replacement. Spare parts are becoming extremely difficult to obtain.

None of the locomotives utilised on Sandite duties are owned by Railtrack and are therefore hired to undertake the work. The costs of running such contracts are proving to be very expensive.

Many of the vehicles are very specialised and stand 'idle' for a large amount of the year thereby making poor use of resources.

Railtrack commissioned The engineering **link** to carry out a feasibility study to establish the practicalities of modifying existing Railtrack and other second-hand vehicles including analysing the space and weight requirements for a special purpose designed vehicle to provide the above services. The study took due cognisance of Railtrack's key requirements listed below, and in particular the desire to introduce a modular exchange feature for such equipment. These requirements were determined by the key personnel within Railtrack's Zones who were responsible for providing the services.

Table 1
Railtrack's Key Requirements

- Preference is for "off the shelf" vehicle(s) or modules having a proven design already in service elsewhere
- Constructed to maximum of W6 loading gauge.
- Must be Route Availability of RA6 or below
- Power unit shall be sufficiently powerful to maintain 60 mph or a 1:200 rising gradient with a full payload.
- Must be capable of being driven from either end.
- Operator(s) must be able to change ends without leaving the vehicles.
- Fluid applications must be speed sensing fitted.
- Must comply with all Railway Group Standards especially crashworthiness and ride index.

- Must comply with all EU environmental standards.
- Fuel range 750 miles
- Water jetting capacity 100 miles
- Sandite application capacity 40 miles (provisional maximum)
- De-icing fluid capacity 800 miles (at current application rates)
- Weed spraying capacity 200 miles
- Availability 90% rising to 95% after two years
- Reliability 50,000 miles per casualty
- To include messing facilities
- To include toilet and washing facilities.
- Snow clearance potential with miniature snowploughs.
- Hot weather fire prevention water spraying and fire fighting capability.

2.1 Method

The study commenced by investigating the services to be provided and the methods currently employed. The space envelopes and weight requirements for the best currently proven equipment was determined. The study then examined possible existing and other potentially available second hand vehicles to ascertain their capability of accepting the equipment in modular form. Consideration was given to the ability of "off the shelf" new specialist Track Renewal and Maintenance Machine (TRAMM) type vehicles to accept this equipment and the feasibility of new vehicles specifically designed to meet the requirements was examined.

2.2 Equipment Requirements

The Geismar High Pressure system had been demonstrated during Autumn 1996 on the SW Zone on the Eastleigh, Salisbury and Exeter route. The water usage was 130 litres/min. which coupled with Railtrack's requirement for 100 miles of track to be treated and a treatment speed of 35 mph, gives the amount of water dispersed as 22,285 litres. The weight of this amount of water is 22.3 tonnes but for operational reasons it was thought prudent to add another 10% for unseen contingencies giving a total of 24.5 tonnes. The weight of the Geismar pumpset package is 5 tonnes and the space envelope required is 5.5m x 3m x 1.6m (length x height x width)

The requirements for the Sandite equipment were based on existing technology although it was recognised that this had its short comings particularly in wheel wear. The weight of the existing typical hopper and discharge equipment is 0.85 tonnes and each Sandite Hopper (two required, one for each rail) is 1.3m x 1.5m x 0.9m (length x height x width). The rate of application of Sandite is 16 litres per mile per rail and the speed of the train when depositing it is 20 mph. Railtrack's requirement was for a capacity sufficient for 40 miles of track to be treated.. Therefore the amount of Sandite used over 40 miles is 640 litres. Sandite weighs 1.625 kg/litre and so the 640 litres is equal to a weight of 0.4 tonnes / rail

Lawrence Industries supply Sandite in 800 litre packs. If this was considered as the charge capacity it would weigh 1.3 tonnes and provide a 50 mile range. This was therefore the recommended amount to be considered for each of the two hoppers. The total Sandite load for the train would therefore be 2.6 tonnes.

For the de-icing system again the existing equipment was considered to determine the sizes coupled with Railtrack's requirements for 800 track miles to be treated at current application rates which are 4.4 litres per track mile. The required capacity was therefore 3,520 litres of Kilfrost which equates to approximately 3.52 tonnes of liquid, say 4.00 tonnes allowing for tank, pump and fittings. The space envelope required on the vehicle would be 3.6m x 1.5m x 1.2m (length x height x width).

To determine the requirements for track bed and trackside weed killing the current package as used by Nomix Chipman was examined. The equipment comprises of three individual items which are:-

(1) Generator - 0.7 tonnes - 1.5m x 1.9m x 1m (length x height x width)
(2) Control Unit - 0.5 tonnes - 2.0m x 1.0m x 1m (length x height x width)
(3) Chemical Unit - 5 tonnes - 4.0m x 2.0m x 2m (length x height x width)

However, the amount of chemical within Item 3 is only sufficient for treatment of 100 track miles and as the Railtrack's requirement called for 200 track miles to be treated the weight would double, to 10 tonnes.

The water usage for weed spraying equates to 500 litres at ½ tonne/mile. For 200 track miles to be treated this equates to 100,000 litres or 100 tonnes and, including an allowance of 10% the capacity to be hauled is 110,000 litres or 110 tonnes. Railtrack's requirement was that the vehicles should meet RA6 or lower. Under this criteria the individual axle weights must lie between 18.3 and 20.3 tonnes For the purpose of this investigation a 19 ton axle load was assumed.

Railtrack's requirement called for the vehicle or combination of vehicles to be self powered and being capable of being driven from both ends. Because of the high capital investment there is an operational need to maximise vehicle utilisation. Therefore, the design concept needed to embrace the requirement that each of the four systems (Sandite, Water Jetting, De-icing, Weed Spraying) must be in modular form so that they may be fitted / removed to provide either of the two different seasonal treatments (i.e. Autumn / Winter or Spring / Summer) within the train formation.

Therefore for ease of mounting and removal of the individual modules clear longitudinal vehicle access between cabs was very desirable. This was essential if the currently proven self contained packages remain in line with the stated space envelopes. Thus the ideal vehicle design needed to be on the basis of a flat platform i.e. a load bearing underframe.

Table 2
Summary of Equipment and Fluid Weights and Space Envelopes for Spring / Summer and Autumn / Winter Operations

Spring / Summer)

Pump Set (based upon Geismar supply)	= 5 t	5.5 x 1.6m (length x width)
Weed Spraying (200 mile range)	= 10.5 t	2 x 1m (length x width)
Water	= 110 t	

Autumn / Winter

Pump Set (water jetting)	= 5 t	5.5 x 1.6m (length x width)
Sandite (per 2 hopper installation)	= 3.45 t	1.3 x 0.9m (2 off)(length x width)
De-icing	= 4 t	3.6 x 1.2m (length x width)
Water	= 24.5 t	

Figure 1 : Existing Railtrack Sandite Vehicles

2.3 Existing Railtrack Owned and Hired Vehicles

From the foregoing, it will be appreciated, from the size of the space envelopes for the various modules plus the volume / mass of water to be hauled for two of the specified functions it would rely on the traction power being provided by a locomotive or motor vehicle. The issue of side vehicular access remains essential for the modular concept to work.

With regards to the locomotives currently used to provide these services, two issues need to be considered. Currently they do not meet the availability and reliability targets specified and it is also probable that as English Welsh and Scottish Railways (EWS) receive their new Class 66 locomotives they may dispose of the older Classes of locomotives which are fitted with Sandite equipment. It is also possible that the other locomotives of similar classes (i.e. Heavy General Repaired (HGR) Class 37's) which they do retain may not be available for these services.

The existing Sandite fitted 2-car Diesel Multiple Units (DMU's) would accept up to about 8 tonnes payload but access to the interior of the vehicles is very difficult for large items. The water jetting would need to be fitted into a separate third vehicle in the middle of the formation plus separate tanks for water carriage. The performance of the DMU power cars with only 600 HP would be insufficient to give an acceptable light running or emergency performance.

The existing Sandite / de-icing fitted 2-car Electrical Multiple Units (EMU's) (see figure 1) would accept a similar payload to the DMU's and again access to the interior of the vehicles is difficult for large items. The addition of water jetting would have to be by additional vehicles as with the DMU's. The performance of two EMU motor coaches with 1000 HP would be sufficient to give an acceptable performance.

2.4 Modification of Second Hand Locomotives

The locomotives considered were Classes 37, 31 and 20. The case for not considering the Class 37's as an option has already been partly stated in paragraph 2.3 above. However, in addition to their age, future spares, reliability and availability, Class 37's have a route availability of either RA5 or RA7. The RA5's are basically those which have not been refurbished and hence first to be scrapped. The locomotives are rated at 1750 HP but could be derated. The ride and cab environments are not to current standards and the locomotives are approaching 40 years of age, i.e. at the end of their design / service lives. The vehicles are of a monocoque structural design and hence not suitable for a modular concept.

The criteria stated for Class 37's is in the main applicable to Class 31's. They have the same age profile and therefore spares will be an increasing problem. Locomotives of this age profile are designed with a monocoque structure. Additionally if used as a non-powered driving vehicle the space available between the cabs would not be long enough.

The Rolling Stock Library advised that there were up to 25 Class 20's on their books. Thus, if these were all made available with 12 used for traction and the other used as a non-powered driving platform, only 12 train formations could be provided. The requirement was for 40 or more trains.

2.5 Modification of Second Generation DMU'S

The majority of modern DMU'S are on lease for significant periods in order to meet Train Operators franchise commitments. The only units off lease are some Class 141 units. The installed horsepower of a unit capable of undertaking the requirements should be in the region of 1200. The current installed horsepower for a 2-car unit Class 141 unit is only 406.

There is not sufficient room within the wheel base to install an engine similar to the Cummins KTS-19-R or Voith T312R transmission etc. to provide the required installed horsepower or to accommodate the extra fuel needed for traction and for the module services against the specified requirements. The Spring / Summer equipment arrangement in terms of its weight would also exceed the vehicles capacity by 5 tonnes. The total number of Class 141's is 38 vehicles, i.e. 19x2-car sets. If all were available there would still not be sufficient to meet the size of fleet envisaged.

2.6 Existing Purpose Built Vehicles

Within Europe, SNCF, SBB and DB have or are intending to purchase vehicles that have been specifically designed to meet their aspirations for rail conditioning and weed spraying but the modular exchange concept may not have been exploited.

It should be noted that the vehicles currently in service or about to enter service in Europe are currently designed to their loading gauge and additionally they may not meet a number of Group Standards which are mandatory for operation in the UK.

The operation of the water jetting, sandite, de-icing and weed spraying in the fixed vehicular situation could be fully assessed to provide the necessary confidence in terms of individual system reliability. The main design workload would be in the provision of a suitable powered driving platform to accommodate the concept of modular exchange plus all the services and controls plus Global Positioning Systems and remote recording of the identified functions.

If one looks at a typical design of, a Track Renewal and Maintenance Machine vehicle (TRAMM) initial concerns would be as follows:-

(1) The cabs probably would not meet Group Standards especially in terms of end loads.
(2) The platform length would not be sufficient to accommodate the two Sandite modules together with the water jetting module at 5.5m long.
(3) The combined installed horsepower of two vehicles would not be sufficient
(4) The curving ability (2-axle vehicles with typically 7m wheelbase) may not be acceptable
(5) The ride may not be acceptable.

A full detailed scrutiny of the design against Group Standards would almost certainly identify other issues that would require modification.

2.7 New Vehicles Based on Existing D.M.U. Design

When considering the possible provision of a vehicle based upon current UK multiple unit technology modified against the requirements it should be firstly noted that the maximum additional load that may be evenly distributed over the total floor area within a monocoque structure is typically 12 tonnes. The load equates to the crush load of passengers and is also limited by the axle design. However, if the 12 tonne loading criteria for the purpose of this evaluation was considered to be acceptable on the basis that the water for the services was hauled in tank wagons there are other issues to be considered.

Another major factor is due to the fact that as already stated these vehicles have been designed as a monocoque structure. The above floor level structure acts as though it was a tube. Hence, it is not possible to significantly increase the size of the door openings. This design feature would not allow the currently large individual modules to be manoeuvred in or out of the vehicles. This, therefore, defeats the modular exchange concept based on the existing module designs. Smaller packages would have to be designed and proven.

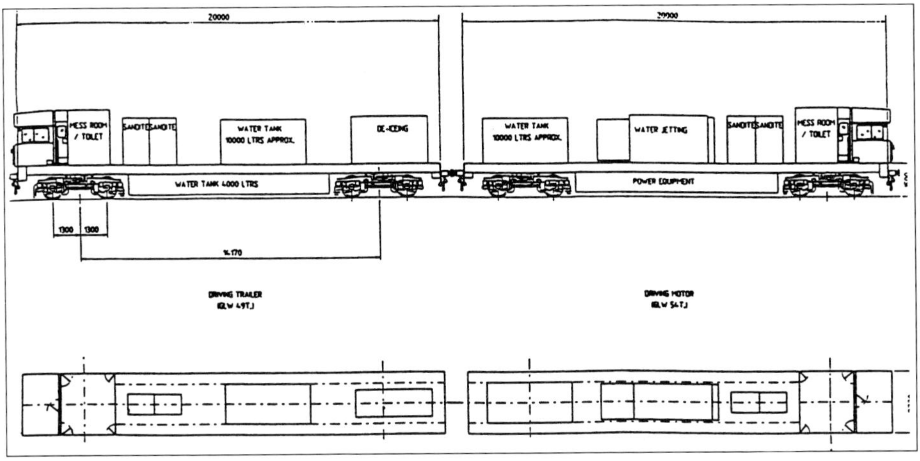

Figure 2 : Proposed New Railhead Conditioning Train

2.8 New Purpose Designed Vehicles

For new purpose designed vehicles based upon a reasonable fleet size there would obviously be up front costs to cover the vehicle design and jigs and fixtures but the final train cost should be lower than that based upon the adaptation of an existing design of passenger carrying vehicle.

An initial conceptual design, (see figure 2) together with life costs was considered by The engineering **link** and the following represents the conclusions.

Each powered vehicle would be need to be approximately 20m long with bogies positioned at 14m pivot centres. The platform height would be approximately 1.5m above rail level. The cab would need to be based upon latest designs to meet the crashworthiness requirements and one of the powered vehicles would be equipped with both toilet and messing facilities in the rear of the cab.

The multi-purpose trains to be employed on seasonal duties are to undertake railhead conditioning by pressure jetting and application of Sandite, conductor rail de-icing and weed spraying. The Railtrack requirements calls for a performance of 60 mph up a 1 in 200 gradient, fully loaded, which equates to approximately 930 HP, 694 kW at the rail. The most arduous of the duties is likely to be weed killing, which, using existing proven technology, will require 110 tonnes of water to be hauled. This, along with the other equipment suggests a train of tank wagons, sandwiched between 2 power cars, each with one underfloor power train to both axles of one bogie, after the fashion of a modern DMU. The maximum train weight would be about 342 tonnes.

A first guess at a suitable power train would be typically the Cummins K19 underfloor engine rated at 600 HP and a Voith 3-speed type 312 transmission.

2.9 Conclusions

The existing Railtrack owned DMU's are not suitable for the addition of water jetting but the existing EMU's could be used as power cars to haul water jetting vehicles. It could only be recommended that the current Railtrack owned DMU Sandite vehicles are retained only for this type of service. Water jetting will need to be provided by either trains propelled by hired in locomotives or by new specialist vehicles.

With regards to second hand vehicles and new vehicles several of these options would require intermediate wagons between the 2 driving vehicles, the number of such intermediate vehicles would be determined by the carrying capacity of the driving vehicles themselves. The base intermediate vehicle was assumed to be a bogie flat wagon with a tare weight of 21 tonnes, a payload capacity of 61 tonnes, and a length of 19.7 metres.

With regards to the existing Railtrack EMU's (DC) it is possible that a trailer vehicle can be inserted between the two motor coaches to provide the water jetting facility.

The availability and reliability of locomotives for hire is a commercial issue between Railtrack and EW&S and would need to be resolved in order to satisfy the short term options or otherwise high damages could continue to be paid to Train Operators as a consequence of poor railhead conditions.

No second hand vehicle options are immediately obvious either due to insufficient number of vehicles available to meet the total requirements or because the vehicles are not suitable for conversion to satisfy the modular concept

With regards to the new vehicle options in particular existing Track Renewal and Maintenance Machine type vehicles, all such vehicles seen to date are 2-axle vehicles and as such a pair are insufficient from a load carrying or from a platform length point of view to transport the equipment needed in the Autumn / Winter season. It is recommended, therefore, that additional powered flat vehicles (uncabbed) will be required unless a bogie vehicle is available.

A conceptual bogie vehicle can be produced in a two vehicle trainset which together can carry the equipment and weight necessary for the Autumn / Winter operation. Comparable vehicles have already entered prototype service in Germany for DB called a Cargo Sprinter manufactured by Windhoff AG and Bombardier Talbot.

Figure 3 : SNCF Water Jetting Vehicle

Figure 4 : Railtrack's prototype Water Jetting Unit

3 EQUIPMENT DEVELOPMENTS

Historically Railtrack has largely relied upon sand-based compounds to address low adhesion, starting with "home brewed" concoctions through to the latest variants of Sandite produced by external suppliers. These have several disadvantages to the railway in that their application is slow (20 mph), they lead to wheel damage on the vehicles applying it and to track components, and increase the risk of wrong side failures of track circuits through the sands insulating properties when dry.

Other solutions used include rail scrubbing and low pressure water jetting. These treatments also have their own disadvantages; rail scrubbing is effective but very slow (walking pace) and low pressure water jetting requires large volumes of water, usually in tankers, and locomotives and other support vehicles to operate it.

Railtrack had identified that the UK was almost unique amongst Northern European countries in the way it addressed rail contamination. Railtrack reviewed European experience, particularly in France which had similar train services and operating conditions to the UK, around Paris for example, and the normal treatment used was high pressure water jetting (up to 1000 Bar - 14,000 psi). In some cases using hot water or additives such as sand (see Fig 3).

Therefore, Railtrack purchased a Geismar high pressure water pump / jetting arrangement to trial during the Autumn of 1996. Trials were conducted in the then Railtrack South Western Zone by AMEC Rail in conjunction with South West Trains on the Waterloo-Exeter route, West of Basingstoke (see figure 4). Railtrack also hired an additional Chipman's low pressure water jetting train for use in the West Anglia Great Northern Railway operating area. The difficulties facing Railtrack were how to measure success as measuring adhesion is a difficult process. The approach taken in 1996 was to measure adhesion by location and over time to identify how sustainable water jetting was. To do this, Railtrack purchased American supplied Portable Tribometers which were advertised as suitable equipment for this task. However, the value of these results was brought into question when the tests on the Portable Tribometers showed them to be inaccurate at measuring absolute low adhesion values. However, other evidence, including anecdotal feedback from drivers and train delay performance data, was encouraging enough to suggest that there was a basis for further work in this area.

Railtrack took this forward for Autumn 1997 based on the experience gained in 1996. The objectives were to overcome the limitations experienced in 1996. Therefore, a new testing methodology was identified; to measure the effects of low adhesion rather than the absolute values, to use service trains with onboard Data Recorders and to select risk-free locations for the tests. This led to the following plan of action:-

- use the North Downs line which is a known problem route
- use a suitable modern disc-braked unit fitted with Train Data Recorders
- enlist the help of a TOC with a real interest in adhesion problems
- ensure enough trains use the route to measure the recontamination effect
- capture relevant micro and macro climate data
- employ experienced and independent consultants to undertake trials

The initial results from the independent consultant, BR Research, demonstrate the importance of Sandite application and that the greatest overall improvement that can be obtained is from a combination of high pressure water jetting and Sandite.

Railtrack also decided early in 1997 that it needed to expand the trials to include de-icing equipment developments and placed an order for 2 VMT 860 PL/UM track motor cars (Track Renewal and Maintenance Machines) from Geismar UK to be fitted with high pressure water jetting, bi-directional Sandite and de-icing equipment. The de-icing development is based upon a sleet brush used by London Transport (see figure 5).

Railtrack have also used the first Geismar pump set during the Summer of 1997 to trial some novel weed spraying techniques developed by Cranfield University.

Figure 5 : Railtrack's Concept Vehicle

4. PRODUCTION VEHICLES

Railtrack issued an Official Journal European Communities (OJEC) notice in late 1996 to cover the procurement and maintenance of an anticipated fleet of new multi-purpose vehicles. A Pre-qualification exercise was undertaken by March 1997 and invitations to tender based on a performance specification, encapsulating Railtrack's requirements, produced by The engineering **link** was issued in May 1997. Tenders were returned by September 1997 and at the time of writing tender evaluation is virtually complete and a business case for part of the total requirement is being put to Railtrack's Investment Panel prior to concluding and awarding a contract.

In parallel in the middle of 1997 Railtrack issued a further (OJEC) notice for the future operation of the multi-purpose vehicles and invited all interested parties to discussion on the possibilities not only for the specified use but also to canvas opinion on other uses. The intention is to have a few regional contracts which will provide a service to Railtrack using Railtrack's own multi-purpose vehicles.

5. ACKNOWLEDGEMENTS

The author would like to acknowledge the assistance given by the following persons and organisations in compiling this paper:-

Railtrack	- S Palmer, M Hunter
Geismar	- D Bambury, J Carré
AMEC Rail	- D Hill-Smith, R Neathey
BR Research	- J Tunley
Halcrow Transmark	- C Fulford
The engineering **link**	- M Frampton

C547/006/98

Reliability – how much is it worth?

N HODGSON BSc, MIEE and **T NICHOLSON** BSc, AMIEE
Booz·Allen & Hamilton Limited, London, UK

SYNOPSIS

The UK rolling stock market is moving rapidly towards the specification of standard modular trains which look to meet the needs of Operators, Lessors and Manufacturers.

However the world-wide rolling stock requirements for performance and reliability vary radically, even for a basic train type, the commuter EMU. This poses the question, how much should I pay for something better or different?

This paper looks at the varying values that cause different railways around the world to be willing to pay for different levels of performance and reliability.

The paper suggests how to specify what you really want for reliability, for there is no point in paying for something you don't need, or can not use.

1 INTRODUCTION

The world-wide rolling stock requirements for performance and reliability vary radically, even for a basic train type, the commuter EMU. An Operator buying new rolling stock is frequently faced with the dilemma of having to define reliability requirements and specify them. Over-specification will result in the rolling stock costing more, yet under-specification could result in the Operator being saddled with major running costs or being forced to fix problems himself.

This paper looks at the values that cause three different railways around the world to be willing to pay for different levels of performance and reliability. Values such as service level, maintenance policy, financial penalties, historical factors and passenger expectations.

The paper concludes by suggesting some methods which will assist Operators in the specification of new train reliability.

2 HOW MUCH RELIABILITY DO YOU NEED WHEN BUYING NEW TRAINS ?

Frequently this question is not asked, or the answer is plucked out of the air "Better than it is now!". But understanding what your requirements for reliability are, and what the risks are if you get it wrong, are fundamental to specifying the right equipment.

And as suppliers are now expected to take greater responsibility for the equipment supplied, often including maintenance, what you specify tends to have a greater impact on first cost than previously.

But as a first step in answering the question, the rules have to be specified. Some of the rules come from the service you plan to operate, and thus require little extra thought for this question, but one new rule is needed, what is a failure?

2.1 What is a failure?

Failures can be split into two basic categories, those which cause trains to be delayed, mission failures, and those which require only maintenance activity to rectify the fault. Both of these can result in an important side effect, reduced availability

It is our belief that non-delaying failures are less critical when viewed against the railways main goal - the movement of people on time. Therefore we will look at service impacting failures, mission failures.

The length of delay caused by a failure will need to be considered. For example, consider a single days operation with two types of failures. One a single failure which lasts for 20 minutes; the other consisting of ten two minute delays spread through the day.

Viewed from the Passengers' perspective the twenty minute delay is likely to be the most severe. It will cause his, and several other trains to arrive late at, and depart from, their destinations. The two minute delays are unlikely to be noticed by the passenger and are unlikely to affect more than one other train on the network.

To avoid the discussion from becoming too abstract we will use three railways to illustrate some of the points we make. While different is some respects, the three railways can be considered to be almost pure "commuter" railways. They move large numbers of people to a single point, an employment centre, with morning peak flows inbound and evening peak flows outbound.

While similar in many respects (at least within the same order of magnitude) it is the differences that are important in the discussion that follows.

3 THE THREE RAILWAYS

3.1 Basic Facts

The three railways that will be considered by the paper will be one in the United Kingdom, one in the Far East and one the United States. Table 1 gives an outline of their relative size, ridership and the service level.

Table 3.1 A comparison of the three chosen railways

	Railway A	Railway B	Railway C
Fleet Size Type	316 cars	351 cars	670 cars
Train length	4 cars	12 cars	3 to 10 cars
Track Length (km)	73	34	130
Number Of Stations	25	13	37
Peak Headway (Mins.)	3.5	3.0	3.75
Passengers per day	60 000	675 000	250 000

3.2 Railway A

Railway A is predominately a commuter railway serving London and was one of the first franchises to be sold off by the UK government via the Office of Passenger Rail Franchising (OPRAF) as part of the privatisation of the its rail network. They are contracted to provide a specified service level for an agreed annual subsidy over a 15 year period, the subsidy reducing annually. Failure to meet the contracted service levels could require payments to be made to both Passengers, under the Passenger Charter and OPRAF.

Railway A is responsible for the operation and light maintenance of their fleet, which they lease from Rolling Stock Leasing Companies (ROSCOs). The ROSCOs are responsible for the heavy maintenance of the units. Railway A are currently replacing their 30+ year old EMUs with a fleet of new 4 car EMUs. These new units will be maintained by the manufacturer, under contract, the end of which coinciding with the end of Railway A's franchise.

Railway A is not responsible for their infrastructure, this being provided by Railtrack under a track access agreement. Included in this agreement is a performance contact which sees Railway A pay Railtrack for delays which are due to failures under its scope, e.g. train failures and Railtrack paying Railway A for failures within its scope, e.g. point failures.

3.3 RAILWAY B

Railway B is wholly owned by the local Government, but operates as a commercial operation, controlled by an appointed board of directors, and operates at a profit, and when property development is included, is essentially self funding.

All services are now run with 12 car EMUs with a frequency of 20 trains per hour peak and 12 trains per hour off peak. The fleet is currently going through a mid-life refurbishment, that is increasing the number of doors per side, and replacing the transverse seating with longitudinal seating, leading to a realistic capacity of each 12 car train of greater than 3 000 people. Railway B has full control over both its rolling stock and infrastructure and is in currently expanding with the addition of a new line.

3.4 RAILWAY C

Railway C is run by a Board of Directors, each of whom is elected by nine local election districts. It controls both its own rolling stock and infrastructure.

Railway C was conceived as a hi-tech railway with automatic train control from its inception. It is also rare in that it uses a non-standard rail gauge. An extension to the railway is currently being constructed. Although in many respects closer to a Heavy Rail Metro, its longest route is currently 47 miles long (75 km) and used almost exclusively by commuters.

4 HOW DO FAILURES AFFECT OUR EXAMPLE RAILWAYS?

The actual effects of a single failure, and of repeated failures can be very different depending upon the railway. These effects are often largely controlled by financial and social effects, rather than simple technical problems.

4.1 Multiple short failures

Firstly consider the effect of periodic failures of duration up to five minutes, that result in the degradation of service over an extended period of time, the effects and impact are very different in each case we are considering.

Many of the passengers who use Railway A have real choices about the way in which they choose to travel. They are able to use other rail operators as well as other modes of transportation, so ridership levels (and thus revenue) could drop and this is of major significance.

Revenue is obtained from two sources, ticket sales and government subsidy. If the passenger revenue reduces it cannot obtain increases in subsidy as this was fixed under the terms of its franchise.

Further, Railway A is committed to providing defined levels and quality of service. Failure to meet these requirements could, ultimately, lead to the termination of its franchise agreement. Thus the avoidance of multiple failures is critical to its success. Thus the avoidance of multiple failures greater than five minutes is critical to its success.

The majority of the passenger who use Railway B are modally constrained, there is little or no viable alternative. A highly controlled and expensive house market limits the opportunities for "moving" away from the problem. The railway is under no formal franchise agreement that requires a level of reliability, and while it operates essentially as a private railway, it does not have the same exposure to financial risk as our UK example does.

Therefore in many respects Railway B is fairly immune to any effects from periodic short duration failures. Being highly motivated to provide an excellent service, Railway B will strive to eradicate the cause of these failures, but as we will see that the problem is unlikely to arise because of the effect of single failures that are considered next.

Railway C is in a different position to both of the other railways. It is a publicly funded operation, and politically controlled. While options do exist to commuting by Railway C, they are generally unattractive; and during a recent strike on Railway C, patrons turned to their automobiles but the increased road traffic resulted in intolerable delays to commuting time.

So the effect of several delays over a week or month to an individual commuter is unlikely to change his mode of travel, and ridership is unlikely to dramatically reduce. Therefore, for Railway C, reliability is driven more by passenger opinion then financial pressure. This effect of this pressure should not be underestimated, the American public are not renowned for being tolerant of poor service!

4.2 Major one off failures

However the picture changes dramatically when we consider single failures, of extended duration.

For a one off failure of say, 20 minutes to an hour Railways C and A are unlikely to suffer greatly. Assuming that the train is stranded out of the station, a majority of the passengers are likely to be seated.

Railway C is in a very similar situation to Railway A. For many people Railway C will be the only public transport part of the journey, and a great many passengers drive to the station.

The economic profile for many of the commuters is high, and flexibility to work from home, tele-commute, is also high. In the event of a major failure, many passengers who are stranded at platforms are quite likely to return home for the day, and accept it as one of those things.

For Railway B the problem is considerably more serious. Given the volume of people on each train (upto 3000 plus in the latest units), of which at least 80% will be standing, and a headway of 3 minutes in the peak, there are often single direction passenger flows of 50,000 to 60,000 per hour during peak times.

Further if the failure affects the air conditioning system then the result could seriously effect passenger health due to high local ambient temperatures for much of the year.

Many of Railway B's passengers typically transfer to a connecting mass transit system and visa versa. This leads to massive back-ups of passengers at key stations when service is disrupted for any length of time. The last major disruption (over an hour) on Railway B illustrates the problem. To control the crowds that built up as a result of the delays, a near riot developed, due solely to the volume of people, and it proved necessary to call out the Army to restore order and control the crowds.

4.3 Non service impacting failure

These are failures that require maintenance attention but do not slow the service down. An example is the loss of an air conditioner, this may result in some discomfort to the passengers in one car and, if very frequent, could result in a decline in ridership.

On each of our three railways, however, the effect of such occasional failures when viewed from a passengers' perspective is negligible. The length of time that any passenger will be on board will be small enough for any failures to be tolerable.

4.4 Reduced availability

Both service impacting failures and non-service impacting failures will require the train to receive the attention of maintenance staff. If the fault is difficult to locate, or requires a long time to remove and replace material then it is possible that the train will not be fixed before it is due to re-enter service. This will lead either to trains being cancelled or "short-formed".

Seen from the eyes of a passenger the problems caused by reduced availability will be identical on all three railways; more crowded trains and extended waiting times.

If, over time the availability of Railway A's leased fleet consistently exceeds that required by service then it has the option of either running additional services or negotiating deals with other train operators for the use of the trains . In both cases profitability will be improved, either through increased revenues or reduced operating costs.

Both Railway C and Railway B own their fleets, with no prospect of selling surplus vehicles. If availability is improved then service could be expanded, but this is often constrained by other infrastructure elements, or a lack of demand. Where availability does become critical is if new vehicles are being purchased, usually for expansion. Here the level of availability that can be relied on determines the number of trains required, and thus has a critical impact on capital cost.

4.5 To sum up the points we have made above

We can see why there are quite different reasons for needing reliability, and different incentives to expend the money to achieve reliability. The frequency and duration of failures that are acceptable will vary from Operator to Operator.

How many failures can you survive? how long is a failure? how critical is it that you make your targets?

As we have seen you may have to be quite specific about the failure requirements, with one rate for short failures, another rate for longer failures, equipment reliability requirements and availability needs. These will all impact the technical solutions required and hence the cost.

Asking for more than you need, say counting two minutes as a failure when five or ten minutes is more reflective of the commercial level for your railway, could result in having to purchase a larger, and hence more expensive, propulsion system.

5 WHAT CONSTRAINS RELIABILITY?

Even if you can clearly identify the level of reliability that is suitable for your operation there may be constraints that will prevent the reliability levels from being achieved.

5.1 Cost

The most obvious and biggest constraint is capital limitation. There is always pressure on first cost, however, if you want the cheapest possible train one of the first things that will suffer will be reliability - regardless of the best efforts of operations and maintenance staff. The effort required to turn a situation around from that sort of start is both difficult and generally lengthy.

Industry capabilities also determine to a certain extent what you can afford. The UK rolling stock market is moving rapidly towards the specification of standard modular trains which look to meet the needs of Operators, Lessors and Manufacturers. Partially as a result of this move prices have fallen, if you have the Standard Product. Moving away from the Standard Product will lead to significant increases in cost.

Railway A concluded that the Standard Product with an associated maintenance contract from the manufacturer represented the best compromise between cost and risk. The maintenance contract includes performance clauses which are very similar to those which Railway A has with OPRAF and Railtrack, thereby passing the risk of failures directly to the manufacturer.

Railway C is constrained because its car is unique, not even using standard gauge. Thus as no other operator uses the vehicle the capital costs will be higher than might otherwise be the case. As Railway C has undertaken, and is undertaking major expansion, there are many calls on the available capital. It also has to compete with the rest of the United States for capital funding from the Federal Transportation Agency (FTA). Railway C therefore has to view the initial capital cost very carefully.

Railway B is acutely aware of the trade-off between capital and operating costs and is primarily seeking to minimise whole of life cost.. Furthermore, Railway B is in a good position to take a long term view of its costs and can borrow money if required, with reasonably good assurance that future revenue will cover costs. Due to the significant impacts of failures, reliability could be worth additional capital, especially if it will allow very high levels of availability to be achieved.

5.2 Train formation

The ability to design in reliability is constrained to the greatest extent by the train formation, and the compatibility requirements with the rest of the fleet. Our sample operators cover the range.

Railway C uses the single car as its lowest divisible service unit. While they usually run with a minimum five car train, and with ten car trains during peak hours, single cars are frequently swapped in and out of consists. Thus each car is essentially a self contained unit and has to be fully compatible with all other cars. This results in the need for complex automatic couplers at each end of every car and due to space limitations, it is impossible to introduce significant redundancy. Thus there is only the road of inherent reliability of components to pursue, as little can be done to remove the problems introduced by having many complex auto couplers.

Further more, even if the "high reliability" car is achieved it will often be running with older, less reliable cars in the same consist, so its impact will be diluted.

The new generation trains for Railway B, on the other hand, will be fixed formation 12 car trains. The train will be semi-permanently coupled and the intention is that sets will only be separated for heavy maintenance. The vast majority of command and control functions will be completed with data networks, and the only interface requirement with the existing stock will be for recovery purposes, essentially just mechanical coupling.

This represents a new clean sheet of paper for the designer. Many areas that have created reliability problems can be completely avoided. Fall back positions can be intelligently designed in, and components, such as propulsion systems, can be sized to ensure single point failures do not cause a mission failure.

Railway A's basic unit is a four-car unit which are built up into eight and twelve cars as demand requires. For their new trains Railway A sees inter-operability with other stock on their route being of lesser importance than the benefits in reliability and functionality that can be achieved using modern train equipment. Like Railway B the trains will be required to mechanically couple for recovery purposes only.

5.3 Safety

Passengers and the public, quite rightly, expect that train travel will be safe. However safety and reliability frequently conflict with one another due to the need to have equipment which ensures that the train is safe. Sometimes there are also specific local safety requirements which need to be incorporated into train design.

In the UK, the dominant safety issue for new rolling stock delivered within the last five years has been signalling compatibility. Trains designs have been subjected to rigorous analysis to demonstrate that they are unable to interfere with Railtrack's signalling equipment. In some cases this analysis has required the installation of Interference Current Monitoring Units (ICMUs) which ensure that the return current does not exceed limits and interfere with the operation of track circuits. It has been a long and difficult process to make reliable ICMUs.

The challenge which faces the manufacturer of Railway A's new trains is to demonstrate to Railtrack they are safe as well as sufficiently reliable.

Railway A's ability to mitigate signalling compatibility by improving its infrastructure is restricted because they do not own it. This is not the case for Railways B and C who have responsibility for both trains and infrastructure. Should analysis show that the best means of achieving signalling compatibility is to make changes to the track signalling, then they are free to do so.

Railway B are currently specifying both new trains and signalling and thus are able to achieve an optimum compromise between train and track.

Railway C is somewhat constrained as it has a significant investment in its existing infrastructure, and automatic operation means that much of the interface has to be identical. But it does control that infrastructure, and over time can balance reliability improvements between track and train.

5.4 Passenger expectations

Passengers are expecting more and more equipment on trains. For example air conditioning, information systems and toilets.

The more equipment that there is the more there is to go wrong, and this will dilute the efforts of maintenance staff. Also much of the new equipment is hi-tech, and will require the services of the best maintenance staff to fix it should it go wrong. This is potentially going to compromise their efforts in fixing the more fundamental parts of the trains should other faults develop at the same time.

6 HOW TO DEFINE AND OBTAIN YOUR RELIABILITY REQUIREMENTS?

As we have shown above how you choose to define your reliability will depend upon the particular railway.

6.1 Modelling

By modelling the trains, their failure modes and the system on which they operate it is possible to see how failures impact upon operation.

6.2 Specification

Write the specification well - if you have gone through the questions we have raised you will understand why you are aiming for a certain level of reliability. You will understand the risks of not achieving reliability levels and you may well have a better idea of what you are able to pay, versus what you could potentially lose.

And only if you understand why you want a certain level of reliability, will you be able to explain it to potential suppliers. Suppliers frequently see figures seeming plucked from the air, and getting no clear explanation of "why?", not unreasonably take the attitude that it is not a "real" requirement, but part of a wish list.

It is common practice to have payments made upon achieving performance and reliability levels. Make sure that this is a two way street, and that there is incentive for the manufacturer to gain. Do not penalise small delays with the same severity as long delays.

Coupled to this is ensuring you can measure the reliability. If your systems are inadequate to identify, and reasonably allocate mission failures, the reliability level specified is irrelevant. That which cannot be measured cannot be enforced.

Find out what the unique features of your Operation are. Define your interface clearly as well as the way in which you operate your trains.

Trains will fail. It is important that when they do the problem is known, understood and repaired quickly. Good diagnostic information is therefore essential, as is managing the data that will come from it.

Determine what your policy on maintenance will be. Will you maintain the trains yourself or will you let a maintenance contract for the trains? If you choose to maintain the trains yourself you will need to pay particular attention to the way on which maintainability issues are to be specified. For example the time taken to identify and replace defective equipment. If you choose to let a maintenance contract with the manufacturer, then this will place the onus onto him, but this will have its price.

6.3 Market and Industry Review

Be realistic about the delivery requirements. While Industry has made real advances in reducing delivery times it is vital that sufficient time is allowed to correctly design and test the design. Shortcuts taken in this area will compromise reliability both in the short term, as well as potentially the long term.

Reliability over time is wholly dependant on the original design, and if this is rushed the cost and trouble of rectifying maters later will be much greater. Reliability is a function of comprehensive design analysis, and this will always take time. Time pressures usually reduce the design cycle long before the manufacturing cycle is cut.

Discuss your requirements with the manufacturers to see if they can be met. Go and see recent new builds to see what lessons can be learnt. Find out where the risk areas lie in the design, what's new what's not worked well in the past.

6.4 Commissioning and Operation
Good failure management buys time. Reliability problems will be encountered in the early days. Good failure management that limits delays to sustainable levels, allows the required improvements to be designed and implemented properly, rather that in an atmosphere of desperation and multiple quick fixes. It is much better to assume you will have failures and plan for that, and be pleasantly surprised, than have it all come as an unexpected shock.

7 CONCLUSION

By the use of our three chosen railways, we have demonstrated that setting reliability levels is a complex issue. What is right for one railway will not necessarily be right for another.

When specifying the reliability requirements for any new trains it is vital that the impact of failures on the particular railway be understood. How many failures can be tolerated, and more importantly, how long a duration can they be. Modelling the train performance on the system on which they operate will assist with understanding the impacts which failures will have. When failures do occur, as they inevitably will, it is essential that they can be diagnosed and rectified as quickly as possible.

There are many constraints on reliability. Issues such as safety, train formation, passenger expectations and most importantly cost can all reduce the ultimate reliability of the train.

Obtaining the reliability from new trains will need sufficient time to design and test the train, the manufacturer to understand why the reliability levels have been set at the levels they have, a performance contract that both incentivises and penalises the supplier. It is however, important to understand that the reliability will not be delivered on day one; a period of time where reliability is built up will be required.

But it is also true that poor initial reliability can no longer be accepted. For a private operator, or an operator running at near capacity the introduction of unreliable new vehicles could result in the collapse of his business.

C547/009/98

Life-cycle costing – far from just a calculation

M CHAMBERLIN BEng, MIMechE
GEC Alsthom Railway Maintenance Services Limited, Birmingham, UK

SYNOPSIS

Privatisation has set many of us with the challenge of not only calculating Life Cycle Costs, but integrating such calculations into practical design solutions and processes to help reduce operating support costs.

Discussing calculation methods and approach with many LCC analysts generates conflicting ideas as to how it should be done, but unless the business is understood, who is to say which is the right way or the wrong way ?

There will be hundreds of different LCC approaches. There has to be differences because we all go about our business in different ways. Tools ranging from home made spread sheets to sophisticated software, commercial approaches and process applications will vary from company to company.

This paper outlines a proactive and practical approach to LCC currently being adopted to address the needs of private maintenance in the rail industry. The approach is far from just a calculation. It is part of a proactive development process starting from pre tendering concept, linking rolling stock design to calculation and involving equipment suppliers, designers and maintainers in partnership roles to help meet the ever increasing demands in containing both train build and support costs.

1. INTRODUCTION

Railway privatisation has set one of the most important and risk laden challenges to many, the containment of rolling stock life cycle costs (LCC).

Who should calculate LCC? Is it just a calculation? What if we get it wrong ? What software? How is it best applied? Only a few of the many questions train builders, maintainers, TOC's and ROSCOS have been faced with since privatisation of the railways. Depending on how each Company approaches its business, different answers will undoubtedly be generated. Conflicting ideas as to the best approach will only really be validated when real and lengthy service experience has been gained.

The start of a project has to take a proactive approach at concept design stage linking design to LCC to help minimise this risk. This paper outlines the LCC and Maintainability approach being made for the rail market to cater for the business needs of GEC ALSTHOM in caring for its product from concept to grave.

2. LCC AND MAINTAINABILITY

The nature of today's business in a privatised railway means that rolling stock could remain the manufacturers assets for part or all of its life. It is essential therefore that LCC be considered and controlled from the concept stage.

Train availability has to be high, both to be competitive and avoid contractual damages for poor performance. Availability is the product of reliability and maintainability. Whilst reliability has always been a responsibility of the rolling stock supplier, maintainability has generally been given little if any attention. Identifying maintainability oversights can be corrected at little if any cost on the drawing board. Oversights identified at build stage may be too expensive to correct and will continue to increase the maintenance costs with consequent effect on the overall LCC.

Engineering resource dedicated to maintainability and LCC applied at the R&D/Production periods bears insignificant cost in comparison to that of the rolling stock. The impact of this resource can dramatically effect operation and support by helping designers and equipment suppliers understand the maintainers requirements at the concept stage and effecting design changes at the drawing board.

The risk of not doing this could result in a much higher operation and support curve which could be very expensive to reduce if not impossible. Identifying maintainability problems once the train has been built is too late. Soaking the necessary resource into the R&D/Production phases shown in Figure 1 can reduce the scale of the operation and support curve.

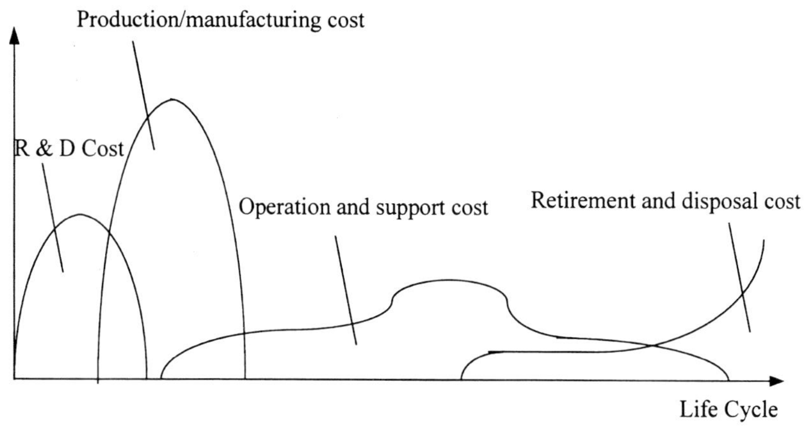

Figure 1. Typical cradle to grave profile. Product should be designed to include the asset LCC.

Maintainability resource applied at the start of a project bears insignificant cost in comparison to that of the asset. It is at this stage that risks need to be identified and rectified to reduce the operating costs. See Figure 2.

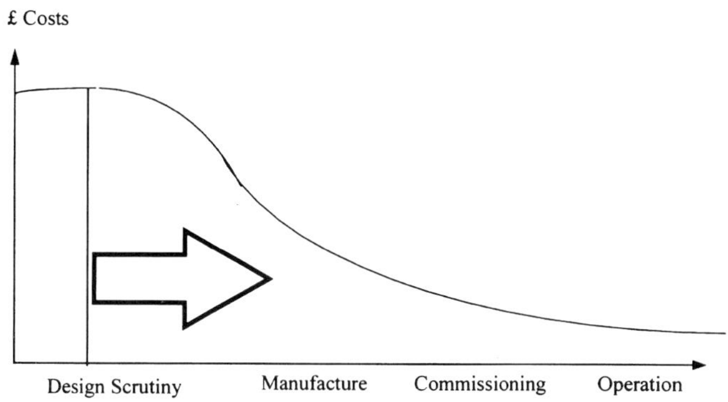

Figure 2. The greatest potential for reducing LCC is during design scrutiny.

The cost of redesign however escalates as the design progresses (Figure 3). Typically :

- Tens of pounds at concept drawing board stage
- Thousands of pounds during research and development stage
- Tens of thousands of pounds during production/manufacturing
- Millions or impossible once equipment is in operation

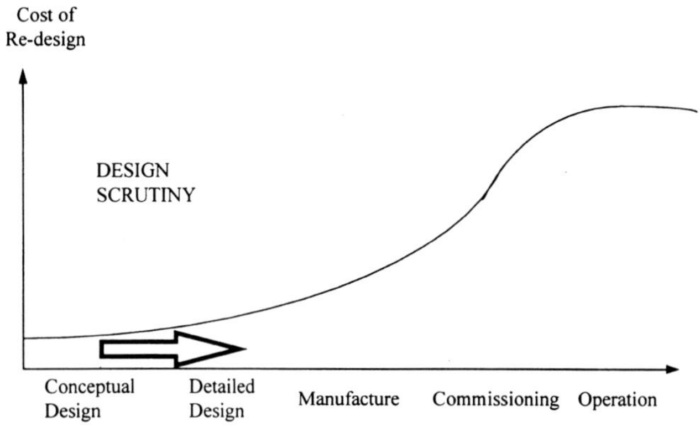

Figure 3. The cost of redesign escalates as the design progresses:

Application of simple maintainability and LCC guide lines is required at the tendering stage to help develop an understanding of the maintainers needs. At this point both the maintainability requirements and LCC have to be considered together as part of the design requirements and monitored through from concept to operating phase to minimise the LCC.

3. THE TOOLS

The key tools required to expedite a process for reducing LCC risk are:

- Maintainability engineers
- LCC modelling and software
- Reliability Centred Maintenance (RCM)

3.1 Maintainability Engineers.

A level of resource with field experience is fundamental to the process. Hands on experience from having shift worked on trains in depot environments needs to support the engineering and design process. Whilst design engineers have extensive design experience, support from maintainability engineers who have actually worked around and below a train, now becomes a fundamental requirement in the design process. Maintenance practicalities can be advised and accommodated in the design from an early stage which otherwise in the past would have been overlooked or ignored.

A set of maintainability requirements forms the starting block for the Maintainability Engineer. The responsibility of this role then needs to ensure that the requirements are applied to the design process. LCC must be linked to these requirements from the start and financial studies made to support design change where there is a reluctance to modify when LCC is being compromised with design issues.

3.2 LCC Modelling.

GEC ALSTHOM have developed an LCC/Maintainability process together with LCC modelling software, tailored to suit their business needs for the rail market.

The LCC model is fed with data inputs provided by train equipment suppliers and builds up a complete train level data base on a supplier by supplier basis. The modelling capacity permits many 'what if' scenarios to be applied to test the sensitivity and impacts upon the maintenance costs. Benefits of RCM conclusions such as maintenance periodicity changes and downtime reductions can be calculated rapidly. LCC modelling with this flexibility provides a powerful tool to help develop an effective maintenance regime.

The model and collation of data is initially controlled by tendering and maintainability engineers with outputs being fed to purchasing, finance and commercial departments. Comparisons are made with the model to determine the most competitive offer not only on capital outlay but LCC as well.

When equipment suppliers are selected, their data is used as the basis of a supplier/maintainer service provision contract.

3.2.1 Data Collation

Data is essentially focused on labour man hours, material cost and task frequency for scheduled maintenance, overhauls and corrective maintenance. Energy consumption and train acquisition costs are currently controlled in parallel to the LCC by the train build authority. These activities may be amalgamated at a later date should the business needs require this. The format in which data is requested from equipment suppliers is shown in figure 4. The LCC model has been developed to operate upon a limited amount of input data fields. This was found necessary to avoid inundating equipment suppliers with requests for large amounts of data. Even at this level, many problems and frustrations have been experienced but after much discussion, help and data iteration this becomes much easier the second time around.

1. SCHEDULED MAINTENANCE

COMPONENT DESCRIPTION			1. Scheduled Maintenance			
Line Replaceable Unit Name and Part Number	Quantity of LRU Per Train	Cost of LRU each (£)	Task Description	Material Annual Cost per Train (£)	Labour Annual Depot Man hrs. per Train	Periodicity Task Frequency Weeks.
BOGIES.						
Motor Bogie Complete	6	22000	14 day visual inspection		13.00	2
Trailer Bogie Complete	6	14000	14 day visual inspection		13.00	2
1A1/01 Motor Bogie Wheelset	12	2800	Tyre condition Visusl		3.25	4
1A2/01 Trailer Bogie Wheelset	12	1950	Gauge Tyres			26
			Profile wheels.	25	28.00	52
Axle box assembly	24	350	Inspect Axlebox/tiebar		3.25	2
1B1/1 Lateral dampers.	12	260	Inspect Lateral Dampers			2
			Inspect Damper Bushes			26
1B2/1 Vertical Damper.	24	170	Inspect Vertical Damper			2
			Inspect Damper Bushes			26
Flange lubrication Sticks	8	7	Fl. Lube sticks.Refill	1456	35	2
Flange lubrication applicators	8	100	Clean		1.00	12
Brake Caliper	48	1200	Examine			
Brake Pad	96	36	Replace	4493	40.00	
Primary suspension springs.Pair.	48	140	Inspect Chevrons			26

2. SCHEDULED OVERHAUL

COMPONENT DESCRIPTION			2. Scheduled Overhaul				
Line Replaceable Unit Name and Part Number	Quantity of LRU Per Train	Cost of LRU each (£)	Task Description	Periodicity Overhaul Task Frequency Years.	Material Cost Per Task Per Train (£)	Labour Remove Man Hrs. Per Train	Refit/Setup Man Hrs. Per Train
BOGIES.							
Motor Bogie Complete	6	22000	BOGIES Remove /Refit	6		20	30
Trailer Bogie Complete	6	14000	Bogies wash	6		12	
			Bogies strip/rebuild	6		144	200
1A1/01 Motor Bogie Wheelset	12	2800	Overhaul. Motor Bogie Wheelset	6	16740		
1A2/01 Trailer Bogie Wheelset	12	1950	Overhaul. Trailer Bogie Wheelset	6	27628		
			NDT Axles	3		48	
Axle box assembly	24	350	Axle box assy. Motor. Overhaul	6	1200	6	6
			Axle box assy. Trailer. Overhaul	6	1200	6	6
1B1/1 Lateral dampers.	12	260	Lateral Dampers Replace	3	3333	12	12
1B2/1 Vertical Damper.	24	170	Vertical Damper Replace	3	4523	12	12
Flange lubrication Sticks	8	7					
Flange lubrication applicators	8	100					
Brake Caliper	48	1200					
Brake Pad	96	36					
Primary suspension springs.Pair.	48	140	Replace.Primary suspension	6	7795	2	4

3. CORRECTIVE MAINTENANCE

COMPONENT DESCRIPTION			3. Failures. Corrective Maintenance.					
Line Replaceable Unit Name and Part Number	Quantity of LRU Per Train	Cost of LRU each (£)	Material Annual Replace Cost Per Train (£)	Annual Repair Cost Per Train (£)	Labour Annual Remove Man Hrs. per Train	Annual Refit Man Hrs. per Train	Annual Diagnosis Man Hrs per Train	F Rate LRU Train Failure Rate FPMK Per Train
BOGIES.								
Motor Bogie Complete	6	22000						
Trailer Bogie Complete	6	14000						
1A1/01 Motor Bogie Wheelset	12	2800		2.08	0.13			0.00861
1A2/01 Trailer Bogie Wheelset	12	1950		3.44	0.17			0.01148
				0.86	0.04			0.00287
Axle box assembly	24	350		0.00	0.00			
				2.25	0.15			0.01148
1B1/1 Lateral dampers.	12	260		124.33	13.78			3.444
				0.00	0.00			
1B2/1 Vertical Damper.	24	170		84.37	6.89			3.444
				0.00	0.00			
Flange lubrication Sticks	8	7		0.00	0.00			
Flange lubrication applicators	8	100		26.00	2.00			2
Brake Caliper	48	1200						
Brake Pad	96	36						
Primary suspension springs.Pair.	48	140		0.36	0.53			0.03444

Figure 4. LCC Data Collation Format.(Data non valid, for demonstration only)

A complete train level LCC data base is built up from many individual train sub system files. This is particularly important where long term supplier partnerships are being considered.

When all the data has been entered into the model it then needs to be driven to reflect the required profile characteristic. The software driving screen shown in Figure 5 allows the LCC analyst to enter for example, the number of trains and annual delivery rate, labour rate, annual mileage, labour efficiency, inflation and discounting rates etc.

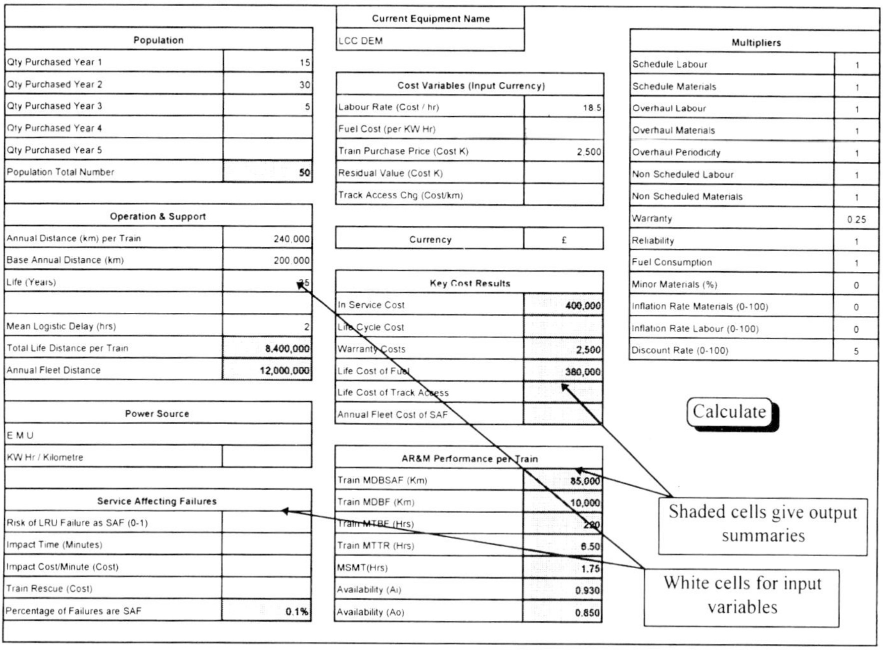

Figure 5. LCC Driving Screen.(Data non valid, for demonstration only)

Once the LCC has been calculated the following are some of the outputs available:

- Whole life cost including asset capital and disposal broken down into material and labour
- Whole life cost Histogram details. See Figure 6.
- Maintenance Star Chart. See Figure 7.
- Material Consumption Spares Listing
- Cost of Failures
- Reliability, Maintainability and Availability Predictions
- Warranty costs
- Service provision contract data compression. Supplier by Supplier LCC targets. See Figure 9.

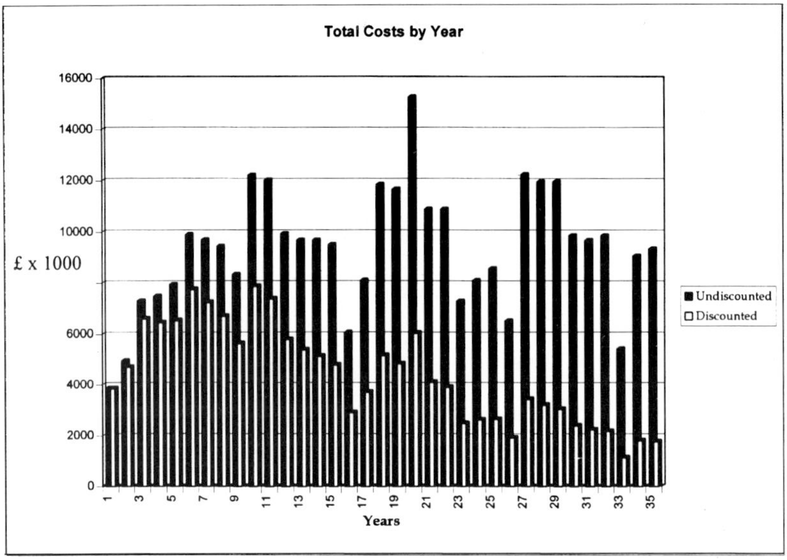

Figure 6. Whole Life Cost Histogram. Each bar can be interrogated to varying detail.

With a model set up many what if scenarios can be tested to determine the sensitivities upon the LCC. For example, what would the LCC change to if the following were amended ? :

- Reliability rate
- Labour rate
- Maintenance/overhaul period (Benefits of RCM can be calculated)
- Number of trains
- Mileage variation
- Labour efficiency
- Inflation rate
- Labour efficiency etc.

At this stage the LCC reflects a conceptual train. It is obvious that things will change as the design develops. Using the base LCC as part of the design development process will help to identify cost drivers so that they can be considered before the hardware is built. Cost benefit analyses can be made with the help of the LCC modelling and used to determine the most cost effective solutions.

As well as this, the maintenance star chart output which summarises task periodicity's (See Figure 7.) is used by the Technical Publications Department. As data flows in from equipment suppliers for the maintenance manuals, the LCC star chart is used as a benchmarking tool to ensure that any deviations from the original LCC data can be investigated. This part of the process helps to confirm that the equipment suppliers engineering and publications organisations are using consistent data, or where design changes have not been notified to the maintenance authority.

Line Replaceable Unit Name and Part Number	1. Scheduled Maintenance Maintenance Task Description	Time per Task / LRU Man Hours	Time per Task / Train Man Hours	Materials Cost per Examination (£)						
				0	0	0	0	68	0	458.1
				Total Man Hours per Examination						
				1	0.1	2	0.1	2.7	3	9.06
				Examination Periodicity in Weeks						
				2	4	8	12	26	40	52
Bogies	Visual inspection	0.75	6.00	*						
		2.00	16.00					*		
		0.00	0.00							
Motor mounted wheelset	Wheel profile checking	1.00	4.00			*				
	Wheel reprofiling	1.50	6.00						*	
	Axle box greasing	0.25	1.00							*
		0.00	0.00							
		0.00	0.00							
Trailer mounted wheelset	Wheel profile and disc checking	1.00	12.00			*				
	Wheel reprofiling	1.50	18.00						*	
	Axle box greasing	0.25	3.00							*
		0.00	0.00							
		0.00	0.00							

Figure 7. Maintenance Star Chart Extract from LCC Model

The LCC forms a part of the offer made to customers at tendering stage. Once the offer has been made the LCC target cannot change albeit the contents can. To share any risk it is necessary to formulate partnerships with equipment suppliers focusing a service provision type contract on the LCC submitted. One of the outputs of the LCC software is a compression of the detailed LCC data into a few lines of information. It is this information which forms the basis of the supplier/maintainer service provision contract. Supplier partnerships are discussed in further detail under section 5.

3.3 Reliability Centred Maintenance (RCM)

Traditionally in the past, the formulation of maintenance schedules has been based on information provided by equipment suppliers. The only problem with this is that periodicity's were set without any real knowledge of how similar equipment's were performing in their operational environments.

RCM is a well proven process, it formulates a maintenance schedule from first principles linking maintenance to design and operation. The process starts from early hazard identification analyses at concept design stage and assesses safety risks from FMECA outputs to ascertain what maintenance is necessary. Unlike FMECA the process starts with a system function rather than at a component level to identify a more general system failure effect. It involves designers, engineers, suppliers, maintainers and operators. The final maintenance schedule and safety analysis can be made at well developed design stages or even after the train is built but the right timing has to be considered as the process can identify redesign issues.

RCM is a live process which needs to be applied from the start and progressed well into the operating phase of a project. The maintenance schedule needs to be continually developed through this process from real train operating experience, until the level of maintenance reflects the failure trends of the train .

Only too often has experience shown that assets operating perfectly well fail after having received maintenance. Too much maintenance can be as damaging as too little. RCM sets about delivering the correct level of maintenance which compliments the challenge to retain LCC.

4. THE PROCESS

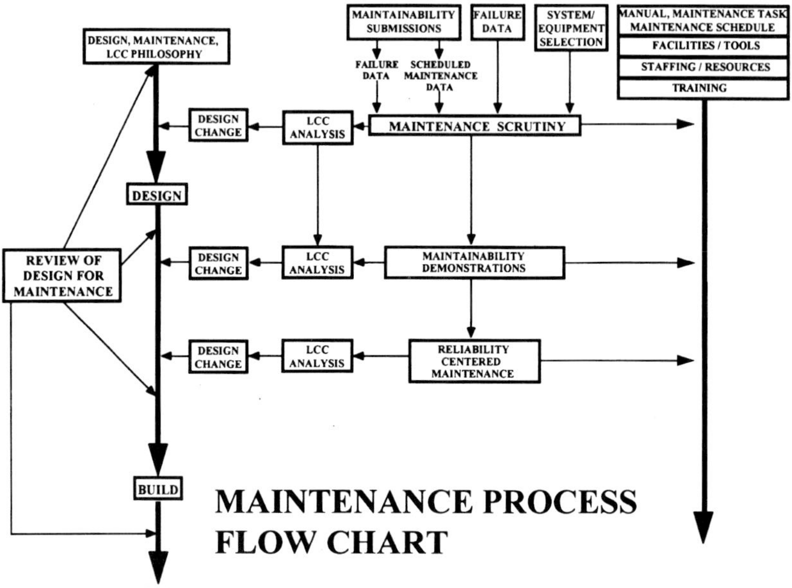

Figure 8. Diagrammatic representation of the LCC control process.

The left hand side of the diagram shown in Figure 8. represents the different project stages of the train design authority. The right hand side is the maintainability and LCC support from the maintenance authority.

LCC data is requested from train equipment suppliers at pre contract stage to assess the most competitive whole life cost (Asset cost + LCC). At the same time maintainability requirements are specified. Information concerning LCC and maintainability will later become contractual. Partnership approaches are then encouraged to develop the design with continual dialogue between the maintainer, suppliers and designer through maintenance scrutiny reviews to ensure that the LCC remains on target.

The maintenance scrutiny approach is applied throughout concept, build and operating stages. The chart above shows a two way process flow. Maintainability engineers feeding experience into the design and design engineers keeping the maintainer aware of design progress. Corrective action can be initiated where any LCC deviations are identified but more importantly as the process begins at the concept stage, design changes can be accommodated at the drawing board with minimal if any cost increase.

Also, from concept stage, FMECA analyses conducted by the train designer are assisted by maintainability engineers. FMECA results are later fed into the RCM analysis described in the previous section, forming a direct link between train design and maintenance.

Maintenance scrutiny looks at the design from a practical view point and flags any issues which can impact on maintenance. Issues which are apparent to the maintainer are not always so straight forward to the designer and likewise design issues are not always appreciated by maintainers. Working as a team helps develop the understanding of the designer and maintainer in one camp. Benefits and concerns can be aired to determine satisfactory compromises. Examples such as:

- Mounting a compressor on the side of the trains underframe instead of at the centre line to enable removal by forklift truck without having to work in a centre pit below the train.
- Redefining axle box clamping locations to prevent the removal of other equipment when clamping the wheelset in an underfloor wheel lathe.
- Changing from brazed to flanged fittings to ease replacement of a water cooling pump.

These are all practical common sense issues which due to engineering design and timescale pressures can easily be overlooked if maintenance is not addressed at the correct time. Maintenance scrutiny commences with the contractual maintainability requirements and the LCC analysis. The purpose of any design review is to confirm and formally log that the requirements are being understood, considered and delivered. LCC analyses have to be made where maintainability requirements deviate from the base, to determine if the financial case is worth pursuing.

As the design develops into the build stage maintenance scrutiny gets easier because the physical asset exists. Design change at this stage however becomes very difficult and it is here where the extent of the maintenance scrutiny benefits should become most apparent. Maintainability demonstrations can begin when the train is in build to validate contractual maintenance targets or generate further change if necessary. Whichever the case the LCC data base needs to be continually revised to reflect the status of the design to be ready for operation.

Maintenance scrutiny and demonstrations should continue up to the operating stage of the project. In particular, when the train is delivered, maintenance demonstrations need to continue at the depot to ensure that the interface between depot, tools and train has been adequately considered. The RCM team continue in parallel to these activities developing the maintenance regime.

5. LCC AND SUPPLIER PARTNERSHIPS

Labour and material costs advised by the equipment suppliers at tendering stage form a risk. Is the data correct? Railway maintenance experience and bench marking will obviously provide some guidance to what these levels of risk are. However, the true LCC will not be known until reliability growth and steady state operation has been reached. Until then, only the LCC and maintainability approach has been in existence from the team working approach between train designer, equipment suppliers and maintainer.

Developing partnerships between maintainer and equipment suppliers is believed to be key to the success of service provision types of contract. The LCC risk needs to be shared, after all why should a maintainer bear the cost of operation if the suppliers LCC data turns out to be well off target. A form of partnership (albeit non contractual at this stage) will already be in existence through the maintainability and maintenance scrutiny approach. This becomes formalised when a contract is formed on the basis of the LCC data which was provided at concept stage.

One of the outputs from the LCC model described previously is a condensed summary of the detailed data base provided by the equipment supplier. Information is condensed into a single line summarising scheduled and corrective maintenance material costs, and man hours labour. See Figure 9.

1. Scheduled Maintenance			2. Scheduled Overhaul					3. Failures. Corrective Maintenance.					
Material	Labour	Periodicity	Task Description	Periodicity	Material	Labour		Material		Labour			F Rate
Annual Cost per Train (£)	Annual Depot Man hrs. per Train	Task Frequency Weeks.		Overhaul Task Frequency Years.	Cost Per Task Per Train (£)	Remove Man Hrs. Per Train	Refit/Setup Man Hrs. Per Train	Annual LRU Scrap/Replace Cost Per Train (£)	Annual LRU Repair cost Per Train (£)	Annual LRU Remove Man Hrs. per Train	Annual LRU Refit/set up Man Hrs. per Train	Annual LRU Diagnosis Man Hrs per Train	LRU Train Failure Rate FPMK Per Train
£ pa	Hours pa		Totals for year 3 overhaul	3	£ Tot.	Hours	Hours	£ pa	£ pa	Hours pa	Hours pa	Hours pa	FPMK Tot.
			Totals for year 6 overhaul	6	£ Tot.	Hours	Hours						

Figure 9. LCC Data Compression. Basis of contract for maintainer/supplier partnership.

A long term contract based on LCC provides benefits to both maintainer and equipment supplier. Maintainers benefit by sharing the risk of poor performance with the supplier of the equipment. The supplier is directly exposed to the operating performance and faces financial risk if performance is not achieved. Performance in an LCC sense is not only reliability but also maintenance and repair downtime and cost of materials. The condensed LCC line above indicates typical performance criteria to be met with a service provision type of contract. Mean time between failure forms only a part of the equation.

A further benefit for the maintainer is the technical support which can be provided by the supplier. Contracting in this way means that there is a great incentive to get it right. Maintainer and supplier partnership and close team working is essential. The key has to be working together to get it right and sharing the problems until resolved satisfactorily. These partnership roles have been under development now for a number of years on some GEC ALSTHOM projects through maintainability studies and RCM forums and are proving to be a valued and enjoyable means of working with equipment suppliers.

Suppliers benefit from a number of things with a service provision type of contract. For example, incentives for exceeding performance expectations, steady long term business in overhaul and repair work, provision of materials and direct and active involvement with their equipment operation. This latter benefit is particularly important. Generally in the past, suppliers have little or no knowledge of how equipment performs once it has fallen outside of the warranty period. Gaining day to day information over much longer periods of time will improve the quality of information in the data bank which in turn will help the supplier in developing and improving the product.

6. CONCLUSION

Life Cycle Costing starts with a very detailed calculation but the process by which it is controlled is far from this.

This paper has described one method of how a proactive approach is being applied by train manufacturer, maintainer and equipment suppliers to help control LCC risk.

There will be many ways in which Companies apply LCC and maintainability, each tailored to their business needs. Operating experience alone will validate the LCC and only at that stage can getting it right be confirmed.

ACKNOWLEDGEMENTS

The author wishes to thank Mutual Consultants Limited and BMT Reliability Consultants Limited for information used for some of the figures in this paper.

C547/030/98

A new lease of life

T DUGHER BSc, CEng, MIMechE
ATC Limited, London, UK
C MORBEY
Adtranz, Derby, UK

THE CHALLENGE

At the start of 1996 the situation for train builders was vastly different to the situation now. The last order for new passenger trains had been placed in October 1993, and more than 2 years had gone by since then.

An analysis of the EMU fleets in service showed the heavy reliance on the ageing Class 42X fleet, although these were not necessarily to be the priority for replacement, and many of the fleets had leases committed past 2000.

The Class 42X vehicles are widely used, versatile and reliabile.

However, they suffer from major safety problems in 2 areas:

Firstly, crashworthiness, the performance of a Mark 1 vehicle in a crash situation is widely known and well documented. The problems are those of over-riding and consequent loss of survival space, due to lack of strength in the upper body and the lack of energy absorption.

Secondly, Slam Doors. The passenger doors on these trains are not interlocked which means they can be opened at any time whether the train is moving, or at a standstill. This can result in passenger injuries, and is of equal concern to the Safety Authorities.

The challenge was to find an attractive, cost-effective solution to solve these safety problems, well in advance of the replacement date for these trains, whilst bringing additional benefits to the operator, and extending the life of the vehicles.

THE CLASSIC CONCEPT

First an existing vehicle would be delivered from service. It would have the cab, guard's brake area, and underframe equipment removed and these would be subject to a C4 overhaul. Then the whole upper body would be removed including the bodysides, end roof and the interior and this would all be scrapped. The underframe and bogies would be retained.

The cab end of the underframe would be cut back to allow the new crashworthy cab to be fitted. Bar couplers would be fitted at the intermediate ends and an autocoupler at the front.

New aluminium bodysides and roof and a steel cab and intermediate ends would then be added to form a monocoque bodyshell. Sliding pocket electro-pneumatic doors could then be fitted.

The vehicle would then be fitted with a new interior and a new exterior livery ready to re-enter service with a new image.

The vehicle would then have a further 15 years life, but because the majority of high value components have been re-used, the cost of the vehicle would be significantly lower than that of a new train.

THE PARTNERSHIP APPROACH

In July 1996, Adtranz launched the Networker Classic concept as being the solution for fast cost effective replacement of the large volumes of Mk1, slam door commuter stock operating in the South of England.

Immediately following the initial launch, Adtranz started talking to rolling stock owners to obtain a Mk1 vehicle on which to prove the Classic concept, with a view to producing a prototype vehicle.

Angel Train Contracts saw this as an opportunity to adopt a partnership approach to this important development project, in order to meet a customer need, and was willing to offer a Class 421 driving trailer as the development vehicle.

Hence detailed discussions began on the partnership approach, and how to tackle the:

- costs of the project
- project scope
- project timescale and milestones and,
- marketing the product.

Having agreed the above, the next step was to agree a project plan, bearing in mind that the development, if it was to succeed, needed to be completed relatively quickly.

THE PROJECT PLAN

The target timescale for the project from start to finish was one year, to include the following:

- Existing vehicle testing
- Design
- Prototype Production
- Safety Approval
- Business Case Development.

The object of this project was to cover not only technical feasibility but also Safety Case and Approval issues, and development of the design for both Classes 421 and 423 to enable swift production commencement following successful completion of the prototype.

Turning now to the vehicle donated for 'Classic' conversion, the vehicle in question is 76112 – this being a Driving Trailer from a Class 421 (4-CIG) EMU, which was built at York between 1964 and 1965.

THE DONOR VEHICLE – HISTORY

The Class 421 and 423 fleets represent a large percentage of the Mk1 fleet still in operation in the South of England (approximately 1300 vehicles), hence the donor vehicle is representative of the target production fleets.

The vehicle is a typical Class 421 vehicle, being the driving trailer from a 750v dc EMU using 1963 camshaft control traction equipment and It consists of a typical Mk1 body mounted on steel underframe sitting on B5 trailer bogies. The majority of Class 421 and all Class 423s have Mk6 motor bogies.
Auxiliary equipment such as lighting, control and brake is supplied by a motor generator set, and the vehicle is braked by the EP/auto system.

The gangway cab ends mean that the driver's cab is necessarily small, and you will see later how the existing control equipment and instrumentation was integrated into a modern ergonomic cab layout.

The interior of the vehicle layout is pretty similar to the way the vehicles were built, with a mixture of low density compartment style and open style seating. The interiors were facelifted in the mid-80s to include new seat moquettes and contemporary surface finishes, and modern fluorescent lighting and P/A system were installed.

INTRODUCTION TO SAFETY APPROVAL

As mentioned earlier, one of the objectives of the project, was to consider all the safety case issues associated with this modification from the drawing board stage, to avoid compromising any infrastructure interfaces.

As with all radical changes to rolling stock, the three key safety areas that need to be considered are:

- Engineering Acceptance
- Route Acceptance
- HMRI Approval.

With this in mind, the design of Classic was structured so as to avoid importing any EMI changes by using the existing traction equipment and auxiliaries, whilst keeping the Kinematic Envelope characteristics constant by ensuring that the weight of Classic is no greater than the original donor vehicle. The latter philosophy also means that using the existing brake equipment does not import any risks.

Ensuring that Group Standards were adhered to was initially something of an area to debate, given that Classic is somewhere between a new train and an existing train.

In order to ensure a smooth path through the acceptance process, meetings were held with both Railtrack and HMRI very early on so that the safety strategy could be debated, understood and accepted.

Both bodies were supportive of the greater benefit and safety of Classic, and were helpful in providing advice to avoid a situation where potential non-compliance could threaten the viability of the project with the loss of the greater benefit.

The philosophy, therefore, was to treat the Classic development as a 'life extension' project and meet Group Standards on new equipment, retain Grandfather Rights on existing equipment, and apply the ALARP Principle to any non-compliances.

Before describing the process for dealing with safety approval, it is worth stressing the substantial global benefit of Classic relative to the Mark 1 vehicle generally, noting that the concept offers a crashworthy bodyshell incorporating energy absorption and over-ride protection whilst eliminating slam doors. More importantly, these benefits can be achieved much earlier than realisable with new trains.

TESTING OF THE EXISTING UNITS'

A key aspect of the engineering design and manufacture would be the successful integration of the existing underframe and equipment, with the new bodyshell parts and equipment.

As much information as possible was needed about the existing units, and although Adtranz was able buy a lot of the original drawings, the actual design data was more difficult to obtain.

It was decided that it would be easier to start in effect with a blank sheet of paper and generate data from the vehicles themselves.

There were 6 types of vehicle that needed testing:

421 DTC, TOSL & MBS
423 DTC, TOSL & MBS .

The tests were carried out in Derby at the Adtranz Engineering Development Test Laboratory.

The following tests were carried out:

Vehicle weighing - each vehicle type was weighed wheel by wheel to establish individual wheel, axle and total vehicle weights in tare, crush and laden conditions.

Bogie Rotational Tests were performed on each vehicle type to demonstrate that they were able to negotiate the designed minimum radius curve, with adequate body to bogie clearance. In addition, the rotational resistance of the bogie/body interface was measured.

Kinematic Envelope Sway Tests. Each vehicle was measured in increments, up to the maximum equivalent canted track inclination to calculate body roll angles, lateral shift vertical drop and body sway.

Solebar deflection measurements were recorded for all vehicles in each load condition.

Vehicle Wheel Off-loading. Each vehicle was tested to determine the effect of wheel off-loading caused by track twist, or some other defect.

Vehicle Centre of Gravity. The centre of Gravity of each of the test vehicles in tare conditions was determined. These tests were carried out without the bogies fitted.

Solebar straightness (plan view) was recorded for each type of vehicle.

Other tests including parking brakes and electrical loading were also carried out.

All these results formed the design parameters for the Classic train.

BUILD PROCESSES

So, enough of concepts and paper designs - the question had still to be resolved - could the concept be brought to life and could it be done for the target price?

The next step was to transform the prototype vehicle to prove it!

The first stage was to remove the equipment that we needed to recover.

The upper body was then cut from the vehicle and removed in a simple operation. This highlights the lack of structural integrity of the upper body.

The underframe was then fully stripped down and shotblasted. It was then subjected to thorough weld integrity tests.

The underframe was then primed and painted, and taken to Derby.

The next question to be resolved was the condition and properties of the undeframe in terms of twist, camber, straightness, corrosion, etc. This was a key issue in determining the technical feasibility of the project.

The cab end was removed and the new headstock welded in place.

The next stage was to weld down the floor sheets and the new floor bearers, replace the pipework and fit the underframe looms into new conduit.

Then the accommodation angle for the bodysides and the supports for the structural partitions, cab back wall and seat rail assemblies had to be welded into place.

Insulation was then added and pre-cut large wooden floor sheets bonded into place.

The underframe module was then complete.
The vehicle was then erected and bolted together. Internal fixtures ensure that it meets the gauging requirements.

The bare bodyshell was then ready for structural test.

In parallel with this programme a separate cab and intermediate end module were progressed.

At the start of the project a full size mock-up of the cab was built and the new equipment layouts done as paper drawings to simulate their positions, and the original cab equipment from the prototype vehicle fitted. Two drivers were invited to comment on this and this gave valuable assistance. This gave the pattern for the whole cab interior. The cab desk was built up as a separate module and it was a challenge fitting in the old equipment and the necessary pipework.

The front of the vehicle is GRP clad. The design was progressed using a 1/5th scale wooden model made to develop the aesthetics.

The requirements of the Disability Discrimination Act and the wish to make the Classic vehicle as friendly as possible to people with a wide range of disabilities is a design priority.

Colour, contrast and tactile buttons as well as audible alarms were used extensively. The wide sliding doors and a suitable ramp make the vehicle accessible to wheelchair users and the design has a toilet module suitable for those with a mobility impairment in the trailer vehicle.

Following test, the bare cab and intermediate end modules were replaced with the two already fitted out.

The vehicle was then fully fitted out: looms were fitted, insulation, the doors, the windows, the interior packages, structural partitions, cab back wall and intermediate end partition, bodyside panels ceiling panels, luggage racks, the carpet, the heaters, seats etc etc.

The vehicle exterior was then painted and the 'Classic' livery applied.

STRUCTURAL TESTS – ALL NEW TESTS

Various sets of tests were carried out alongside the build programme to verify design calculations, such as fire tests on samples of the floor, and the complete modelling of the electrical system on the Systems Integration Rig.

However, the most critical tests which were to prove the Classic concept were the bodyshell static structural tests. The Static load test requirement were taken from Railway Group Standard GM RT 2100.

The bodyshell was tested in the Adtranz Engineering Development Laboratory and took about 3 weeks.

A detailed finite element analysis exercise carried out as part of the bodyshell design process had given confidence that the bodyshell design would be acceptable, however, this had to be confirmed by the testing programme.

There were 14 static structural tests that had to be passed to demonstrate the bodyshell could withstand the specified load conditions without permanent deformation and inadmissable strains.

Over 200 strain gauges were attached to the bodyshell, at critical locations. The bodyshell was then positioned in the test area and supported at the centre pivot locations. Displacement was measured using potentiometers.

Loads were applied using load transducers and combinations of different purpose built equipment to simulate the various conditions.

Examples of the tests are:

Coupler tension, coupler compression, overrriding cases, compression at different positions on the bodyends, and combination tests.

The tests were carried out with the most onerous first, and I am pleased to say the bodyshell passed with flying colours.

SAFETY APPROVAL

Earlier, the general philosophy being applied to gaining Safety Approval of Classic was mentioned, enlarging a little on some key specific areas and update on progress;

- Gauging – The Kinematic Envelope of modified vehicles will not exceed the Kinematic Envelope swept by the vehicles prior to modification.

- Braking – Classic will not exceed the stopping distances presently achieved by existing vehicles. As the existing braking system is retained, the crush laden mass of Classic is controlled by careful design layout.

- EE & CS Compatibility – The existing traction system and auxiliary Power Supply are retained,limiting electrical loadings to within pre-modified levels to ensure infrastructure compatibility is not degraded.

- Vehicle Body – The body structure is compliant with Group Standards as far as is reasonably practicable, and all interior fittings are designed using latest codes of practice.

As far as Railtack approval is concerned, the original Project Safety Plan was submitted and noted by the Systems Engineering SAP in December 1996. Following changes to the Route Acceptance process, the Project Safety Strategy was submitted to and accepted by RSAB in August 1997.

Design Scrutiny is now complete, and RSAB have now granted a Stage D Design Endorsement.

As far as Statutory Approval is concerned, HMRI gave a 'letter of comfort' in June 1996 offering no objections to the proposed development of Classic. Following detailed discussions throughout the development phase, and having studied the Project Safety Strategy, HMRI gave a Letter of Approval in October 1997 to enable vehicle testing to take place as part of the final proof of safety, subject, of course, to operational limitations imposed by Railtrack.

SALES AND MARKETING

When ATC and Adtranz commenced this project, there was a belief that Classic represented a solution to a problem which has been concentrating the minds of various parties for some years - the early replacement of Mark 1 slam-door stock.

However, Classic will only ever be seen in passenger service if it is perceived by the Train Operator, as the right product at the right price.

Given the scope of the project, the market is primarily aimed at the operators of Classes 421, 422 and 423.

The benefits of Classic are many. Apart from introducing a crashworthy body with sliding doors to combat the two biggest safety problems, Classic offers the passenger other benefits such as:

- Modern interior
- More seats
- Wheelchair area
- Toilet suitable for the mobility impaired
- Passenger alarm system
- Modern exterior
- Destination indicator.

It is a mistake to view the Classic as an alternative to the new train, it actually compliments the new train in a commuter operating environment.

Purchasing new, sophisticated trains at relatively high cost to do peak only services is not economic. Classic offers the opportunity to lease a low cost peak only train, to work alongside the new train which can work all-day diagrams.

Another big benefit of Classic, and it goes back to the project scope at the outset, is the ability to enter production very quickly and allow the operator – and therefore passengers – to realise the benefits of Classic much sooner than the 'replace with new' philosophy would allow.

Concentration has, to date, been placed on marketing the vehicle on the former BR Southern Region through various awareness discussions and meetings with potential customers and opinion formers, as the Mk1 vehicle problem is most prevalent here. In fact, there may be other opportunities in other areas.

By far the best marketing exercise has been Railtex at Wembley in November 1997.

It was decided early in 1997 that completion of the Classic prototype for the exhibition was a must, and I am sure many will agree that it was superbly turned out.

Classic has been given quite a lot of publicity through the Railway Media, but it was still surprising that so many people were amazed at the transformation from the original Class 421.

As far as the future is concerned, discussions are taking place with potential customers who are looking closely at the benefits and economics, and the various unit formations which are possible with Classic, to determine the optimum solution and Business Case.

It will be possible to start production of Classic within 6 months of contract signing, and produce for example 30 x 4 car Classics within 12 months of first unit input to give these vehicles "a new lease of life".

C547/012/98

New trains for the new Lantau and airport railway

M W GARDNER BEng, CEng, MIMechE, MHKIE, MCIT, **E N HARRISON,** and
D A GUYLER CEng, MIMechE
Mass Transit Railway Corporation, Kowloon, Hong Kong

1 INTRODUCTION

The decision to build a new self contained railway line between the densely populated areas of Hong Kong and Kowloon to the new international airport at Chek Lap Kok presented the Mass Transit Railway Corporation with a new market - the international airline passenger complete with luggage. However, the new line had also to provide a commuter service to the new town at Tung Chung as well as much needed relief of the existing Tsuen Wan Line in the peak hours. Construction costs would not permit the building of two separate parallel lines facilitating two independent services operating rolling stock of different capabilities and therefore the two services, catering for very different markets, inter-run and largely share common infrastructure and rolling stock technology but still offer significantly different features. The paper describes the market and service requirements, the resulting rolling stock specification, and the procurement, design, and commissioning strategy.

2 ROUTE

Figure 1 shows the 34km route connecting Hong Kong, Kowloon and Tsing Yi island to the new airport and the adjacent new town on Lantau island. There are 4 stations on the Airport Express Line (AEL) and 6 stations on the Tung Chung Line (TCL). Where the Airport and Tung Chung Lines share common stations they have separate platforms and passenger entrance/exit arrangements to maintain complete service segregation. The airport stations have full interchange facilities with taxis, buses and private cars and all stations offer facilities for the physically handicapped and vision impaired. The majority of the route is 2 track except through common station sites where 4 track layouts at upper and lower level are used. Approximately

7km of the route is in tunnel, 1.8km is in stations, 17.6km is at grade, and 8.8km is on elevated concrete structures or on viaducts over sea crossings. It should be noted that Lai King station provides cross platform interchange between the TCL and the existing Tsuen Wan Line (TWL) so allowing passengers destined for the business area of Hong Kong Central (85,000/hr at peak times) a speedier, less congested journey and at the same time reducing congestion on the TWL trains, thus allowing them to pick up more passengers at the remaining 11 stations on their journey to Hong Kong Central. The railway is connected to the TWL by a single non-electrified track allowing the interchange of maintenance trains.

3 SERVICE DESCRIPTIONS AND TRAIN CONSISTS

3.1 AEL

This service is dedicated to the needs of the airline passenger and his luggage and within the total railway journey provides features commensurate with airline travel in terms of baggage handling/stowage, seating, communications, and station facilities. The trains are required to provide an image to both give the AEL service a separate identity and relate the service to the MTR corporate identity. Hence after a process of mock-ups and ergonomic trials the passenger environment was specified as an all seated business class interior (64 seats/car) equipped with baggage racks, under seat baggage storage, seat back TVs and electronic Dynamic Route Maps supported by a public address system. Two wheelchair positions are provided at each end of the train. To facilitate passengers boarding/alighting with bulky baggage and to meet station dwell times the cars have two bi-parting plug sliding doors per bodyside with four baggage racks in each of the two vestibules. Seating is arranged to allow passengers to sit near their stowed baggage and to keep it in view thereby reducing a source of anxiety about the security of their belongings and thus encourage usage of the baggage stowage areas. Similarly a combination of ergonomic trials (Train Operator and passenger detrainment), trial equipment layouts, and structural considerations along with development of the external image of the train front resulted in the building of a full scale cab mock-up. In addition the AEL will offer an 'In Town Check In' service for passengers departing Hong Kong via the Airport which led to the requirement for a special baggage car equipped with fully automated containerized baggage handling equipment.

The initial passenger demand is currently anticipated to be 44500 passengers/day growing to 82200 in 2011. In degraded operation the customer service target is to transport the passenger to the airport within one hour of a train failure which requires that equipment on the train have redundancy in critical areas and in addition the railway has several refuge sidings to allow severely handicapped trains to transfer passengers and baggage to a following train with minimum impact on the passenger or the overall service pattern. The service pattern details are given in Table 1.

Table 1 AEL Service Pattern

	Journey Time to Airport (minutes)	Journey Time from Airport (minutes)	Dwell time (minutes)	Initial Headway (minutes)	Final Headway (minutes)
Hong Kong	23.7	23.6	3 (terminus)	8	4.5
Kowloon	19.3	19.2	2	8	4.5
Tsing Yi	11.9	11.85	2	8	4.5
Airport	-	-	2 (plus 2.5 in turnaround/ cleaning platform)	-	-

3.2 TCL

This line provides a mass transit commuter service of the same concept as the existing MTR lines (i.e. 375 passengers/car crush laden with 48 seated). However the trains incorporate certain enhanced features (air-conditioning, lighting, seating improvements, Dynamic Line diagrams, noise attenuation and an interior design in accordance with the new Corporate identity). Again full scale mock-ups were built to evaluate these improvements and the same cab design was adopted to allow full Train Operator interchangeability between the two fleets. The more onerous emergency detrainment requirement for the TCL trains (2500 people in 28 minutes through the centre front detrainment door of one cab) has dictated the cab configuration for both trains. The initial passenger demand is currently anticipated to be 286000 passengers/day (27,000 at peak hour between 08:00 and 09:00hrs) rising to 491,000 in 2011. The service pattern details are given in Table 2 and the trains must meet the same customer service targets as AEL.

Table 2 TCL Service Pattern

	Journey Time to Tung Chung (minutes)	Journey Time from Tung Chung (minutes)	Dwell time (seconds)	Initial Headway (minutes) Av	Final Headway (minutes)
Hong Kong	22.4	22.45	90 (terminus)	4	2.25
Kowloon	19.45	19.5	30	4	2.3
Olympic	17.3	17.4	30	4	2.3
Lai King	13.4	13.45	45 (cross platform interchange)	4	2.3
Tsing Yi	10.65	10.7	30 (270 when a turn round station)	4	2.3
Tung Chung	-	-	150 (terminus)	8	2.3

It should be noted that the Lai King to Hong Kong journey time of 8.5 minutes compares with 22 minutes on the existing Tsuen Wan Line and that the train performance has been specified to allow for the addition of two future stations on Lantau Island and still maintain the same headway.

In addition to reducing headways to cope with growing passenger demand each train is designed, if necessary, to be lengthened by one intermediate car at a time, with a maximum of eight cars for the TCL service and ten cars for the AEL. At the time of contract award the AEL service was to have started with a 5 car consist and the TCL service with a 6 car consist. Early in the design phase, following modifications to the ridership forecasts, options in the contract for extra cars were taken up and now both services will start with 7 car consists each made up of 3 semi-permanently coupled multiple units coupled to each other by automatic coupler/gangways. Figure 2 shows the development of each train type upto its maximum planned length.

4 DEGRADED OR EMERGENCY OPERATIONS

To support the customer service targets the trains have operational facilities, redundancy, or design capacity to accommodate the following events:

The main degraded or emergency operations are:

- Emergency push out of a failed train
- Emergency passenger detrainment/transfer
- Failed power supply to train auxiliaries
- Failure of an individual door or door system
- Failure of secondary air suspension
- Failure of an inter unit automatic coupler
- Fire
- Passenger Alarm
- Brake degradation
- Pantograph Collision
- ATC Failure

5 INFRASTRUCTURE AND ENVIRONMENT

5.1

The new line is served by 13 substations providing a nominal 1500 volt DC overhead line power supply. Track gauge is standard 1432mm running on UIC 90 rails supported on ballasted track on concrete sleepers and on Sonneville track in tunnels and on elevated concrete structures. Maximum axle load is 17 tonnes. The Automatic Train Control System is a transmission based system that uses a 'distance to go' principle. The train safety is under the supervision of Automatic Train Protection (ATP) with regulation provided by Automatic Train Operation (ATO). Train location detection is by jointless audio frequency track circuits. The ATC intelligence is placed in the on board dual redundant computers, with regular updates of route profile and performance requirement taking place at each platform. At certain platforms facilities exist to download maintenance information from the on board train systems into the railway SCADA system. All stations are equipped with platform screen doors and it is an operating requirement that AEL and TCL trains should, in degraded service conditions, be able to use each others platforms to disembark passengers.

The average temperature in Hong Kong ranges from 0 to 36°C and the average humidity from 60 to 87% and the salinity, particularly on the Lantau Island section, is severe. There is an average annual rainfall of 2200mm but with a maximum intensity of 320mm/h.

To develop the vehicle exterior profile and to maximise the available interior floor area within the constraints of the defined structure gauge and the station environment with platform screen doors, kinematic envelope calculations were carried out at the specification stage using assumed suspension parameters. The resulting vehicle profile was fed into the exterior and interior design studies and verification of the kinematic envelope was ultimately carried out using final suspension characteristics and body profile.

6 MAINTENANCE OPPORTUNITIES

Figures 3 and 4 summarise the planned service regime showing the disposition of trains in service and under planned maintenance. It should be noted that AEL runs at constant headway throughout the majority of the 19 hour service day and the rationale is to run the AEL "standby train" in service is to generate train time in the airport turn round sidings so as to give every train eight minutes for cleaning every round trip. Cleaning provision will be suspended in the event of a train being withdrawn, until another can be despatched from the depot. This is not the case for the TCL where the peak service demand duration will not allow time to both withdraw and introduce a replacement train and off-peak a train is withdrawn from service to stay in the depot as a "hot" standby. It should be noted that the trains were specified concurrently with the design of the new depot dedicated to this line, thus permitting the integration of facilities within the depot and features in the vehicle.

7 FLEET SIZE

The initial passenger demand, service patterns and the planned reliability and maintenance regimes given in 3 and 6 above dictate the minimum fleet sizes for service start up e.g. the AEL initial fleet is 11 seven car trains and the TCL will have 12 seven car trains.

8 PERFORMANCE

The concurrent service pattern demands identical train performance from both AEL and TCL. Table 3 gives the specified performance and Table 4 defines the design load cases.

Table 3 AEL and TCL Train Performance Data

Speed	135 km/h
Acceleration	1 m/s^2
Service braking rate	1.1 m/s^2 (maximum) 0.8 m/s^2 average
Emergency braking rate (two stage rated)	1.35 m/s^2 (1.1 m/s^2 above 90km/h)
Jerk Rate	0.8m/s^3

Table 4 AEL and TCL Load Cases (per car)

Train Type	Tare	Laden	Fully Crush	Crush
AEL (passenger weight = 70kg plus 21kg baggage each person)	0	74 passengers and baggage	114 passengers and baggage	148 passengers and baggage
TCL (passenger weight = 60kg)	0	200	300	375

With a laden AEL train (TCL - fully laden) journey times include a 7.5% coasting margin which reduces to zero in the crush laden load case. Additionally it is an operational requirement that a tare train be capable of pushing a fully laden train of the same type (with its parking brakes applied) a distance of 2km (including gradients of 3%) upto a maximum speed of 22 km/h.

Preliminary performance calculations showed that the railway service pattern would be a high consumer of energy compared with the existing MTRC railway network. To reduce this ongoing cost the regenerative braking capability has been specified to be maximised with minimal use of the mechanical braking system.

9 ROLLING STOCK SPECIFICATION, PROCUREMENT, AND COMMISSIONING

9.1 Specification

The start up fleet sizes are 77 AEL cars (of which 11 are cab cars and 11 are cab baggage cars) and 84 TCL cars (of which 24 are cab cars). To procure separate fleets for each service would have lead to reduced economies of scale in purchase, maintenance, and training costs therefore the trains were specified in a single document with only those areas requiring different features being given special sections or clauses.

The specification was a hybrid being highly prescriptive in those areas of image, methods of operation, key dimensions and layout of both the interiors and the driving cabs, the performance of communication, bodyshell, bogies, traction, braking, and air-conditioning equipments whilst allowing latitude in the technologies to be offered in bodyshell, bogie, traction and auxiliary equipments. Interfaces with fixed equipment were also specified. The performance requirements in the areas of safety, reliability, maintainability, fire, EMC, and noise required the contractor to adopt a design methodology for not only submitting all the appropriate reports concerning each of these areas in parallel with the design submission programme but also to conduct major design reviews with the customer where the compliance of the totality of the train design with the Specification was made clear and any emerging compromises were discussed and accommodated if possible. This process was maintained throughout the concept and detail design phases and on into type testing in the contractors and sub-contractors facilities culminating in the type testing of the train in Hong Kong.

9.2 Procurement

The procurement process followed the MTR procedure of prequalifying companies who responded to advertisements inviting expressions of interest. Pre-qualified companies were issued Invitations to Tender including the full specification and given three months to submit offers. Tender evaluation, bid clarification, and final agreement on the Contract took three months at which point an Option letter was issued to the selected tenderer, a Joint Venture of ADTranz in Germany and Construcciones Y Auxiliar de Ferrocarriles S.A. (CAF) in Spain known as ACJV. Contract commencement could not proceed until the whole Airport Railway project was agreed by the Hong Kong and Chinese Governments some 11 months later (November 1994) which then fixed the opening of the new railway in June 1998 and required the trains to be delivered to Hongkong and to be sufficiently commissioned to take part in the integrated commissioning with all other railway sub-systems by early 1998.

9.3 Commissioning

Apart from the conventional aspects of type testing a selected train and routine commissioning the fleet, the trains are a central part of the process of commissioning a new railway including platform screen doors, Automatic Train Control (which also controls data for the automatic public address announcements and dynamic route maps and carries the interface controls with the platform screen doors), radio (operational communications and data for Seat back TVs and passenger announcements), platform mounted CCTV (providing platform images in the operator's cab) and automatic baggage handling equipment, as well as proving interfaces with power supplies and overhead line. These commissioning exercises are necessary to prove the functionality and performance of the train, all its interfaces, and also include system wide noise measurements, EMC testing, train and service pattern performance (including dwell times and baggage loading and unloading times). The commissioned trains will then take part in operational exercises including degraded operation, evacuations, and certain emergency and fire scenarios.

To achieve train commissioning prior to the readiness of the complete railway a bi-directionally signalled 7 km section of the railway adjacent to the new depot was prepared to allow maximum speed running as well as having a 2km test track in the depot allowing speeds of up to 80km/h with full ATO signalling functions. One three car multiple unit was also taken to the Adtranz test track at Hennigsdorf near Berlin for initial trials on traction and braking hardware and software.

10 DESCRIPTION OF ROLLING STOCK

10.1 Common Features

Bodyshell

The bodyshell structure is 22.5m long measured over the couplers and is fabricated from continuously welded aluminium extrusions with cab cars having, in addition, a welded stainless steel cab frame clad in a glass reinforced phenolic moulding. The bodyshells are painted with the AEL and TCL cars having different liveries to differentiate between the two lines. The bodyshell structure will withstand a longitudinal compressive force of 1500KN or tensile force of 1000KN distributed over the coupler mounting faces in combination with the most adverse vertical loading condition. Anticlimbers are fitted on each body end and the draw gear incorporates emergency energy absorption features. In addition the saloon floor is constructed to give a 30 minute fire rating.

Bogies

The bogies are fabricated from steel with wheel face mounted disc brakes. Motor bogies are fitted with frame mounted three phase motors coupled by a flexible drive to an aluminum cased single reduction gearbox. The bogies have a radial arm primary suspension with coil springs and hydraulic dampers. Secondary suspension is by airsprings supplemented by anti-roll bars. Yaw Dampers are fitted to maintain stability at the higher speeds. A single design of bogie frame serves both fleets of vehicles and both motored and trailer bogies.

With a laden AEL train (TCL - fully laden) journey times include a 7.5% coasting margin which reduces to zero in the crush laden load case. Additionally it is an operational requirement that a tare train be capable of pushing a fully laden train of the same type (with its parking brakes applied) a distance of 2km (including gradients of 3%) upto a maximum speed of 22 km/h.

Preliminary performance calculations showed that the railway service pattern would be a high consumer of energy compared with the existing MTRC railway network. To reduce this ongoing cost the regenerative braking capability has been specified to be maximised with minimal use of the mechanical braking system.

9 ROLLING STOCK SPECIFICATION, PROCUREMENT, AND COMMISSIONING

9.1 Specification

The start up fleet sizes are 77 AEL cars (of which 11 are cab cars and 11 are cab baggage cars) and 84 TCL cars (of which 24 are cab cars). To procure separate fleets for each service would have lead to reduced economies of scale in purchase, maintenance, and training costs therefore the trains were specified in a single document with only those areas requiring different features being given special sections or clauses.

The specification was a hybrid being highly prescriptive in those areas of image, methods of operation, key dimensions and layout of both the interiors and the driving cabs, the performance of communication, bodyshell, bogies, traction, braking, and air-conditioning equipments whilst allowing latitude in the technologies to be offered in bodyshell, bogie, traction and auxiliary equipments. Interfaces with fixed equipment were also specified. The performance requirements in the areas of safety, reliability, maintainability, fire, EMC, and noise required the contractor to adopt a design methodology for not only submitting all the appropriate reports concerning each of these areas in parallel with the design submission programme but also to conduct major design reviews with the customer where the compliance of the totality of the train design with the Specification was made clear and any emerging compromises were discussed and accommodated if possible. This process was maintained throughout the concept and detail design phases and on into type testing in the contractors and sub-contractors facilities culminating in the type testing of the train in Hong Kong.

9.2 Procurement

The procurement process followed the MTR procedure of prequalifying companies who responded to advertisements inviting expressions of interest. Pre-qualified companies were issued Invitations to Tender including the full specification and given three months to submit offers. Tender evaluation, bid clarification, and final agreement on the Contract took three months at which point an Option letter was issued to the selected tenderer, a Joint Venture of ADTranz in Germany and Construcciones Y Auxiliar de Ferrocarriles S.A. (CAF) in Spain known as ACJV. Contract commencement could not proceed until the whole Airport Railway project was agreed by the Hong Kong and Chinese Governments some 11 months later (November 1994) which then fixed the opening of the new railway in June 1998 and required the trains to be delivered to Hongkong and to be sufficiently commissioned to take part in the integrated commissioning with all other railway sub-systems by early 1998.

9.3 Commissioning

Apart from the conventional aspects of type testing a selected train and routine commissioning the fleet, the trains are a central part of the process of commissioning a new railway including platform screen doors, Automatic Train Control (which also controls data for the automatic public address announcements and dynamic route maps and carries the interface controls with the platform screen doors), radio (operational communications and data for Seat back TVs and passenger announcements), platform mounted CCTV (providing platform images in the operator's cab) and automatic baggage handling equipment, as well as proving interfaces with power supplies and overhead line. These commissioning exercises are necessary to prove the functionality and performance of the train, all its interfaces, and also include system wide noise measurements, EMC testing, train and service pattern performance (including dwell times and baggage loading and unloading times). The commissioned trains will then take part in operational exercises including degraded operation, evacuations, and certain emergency and fire scenarios.

To achieve train commissioning prior to the readiness of the complete railway a bi-directionally signalled 7 km section of the railway adjacent to the new depot was prepared to allow maximum speed running as well as having a 2km test track in the depot allowing speeds of up to 80km/h with full ATO signalling functions. One three car multiple unit was also taken to the Adtranz test track at Hennigsdorf near Berlin for initial trials on traction and braking hardware and software.

10 DESCRIPTION OF ROLLING STOCK

10.1 Common Features

Bodyshell

The bodyshell structure is 22.5m long measured over the couplers and is fabricated from continuously welded aluminium extrusions with cab cars having, in addition, a welded stainless steel cab frame clad in a glass reinforced phenolic moulding. The bodyshells are painted with the AEL and TCL cars having different liveries to differentiate between the two lines. The bodyshell structure will withstand a longitudinal compressive force of 1500KN or tensile force of 1000KN distributed over the coupler mounting faces in combination with the most adverse vertical loading condition. Anticlimbers are fitted on each body end and the draw gear incorporates emergency energy absorption features. In addition the saloon floor is constructed to give a 30 minute fire rating.

Bogies

The bogies are fabricated from steel with wheel face mounted disc brakes. Motor bogies are fitted with frame mounted three phase motors coupled by a flexible drive to an aluminum cased single reduction gearbox. The bogies have a radial arm primary suspension with coil springs and hydraulic dampers. Secondary suspension is by airsprings supplemented by anti-roll bars. Yaw Dampers are fitted to maintain stability at the higher speeds. A single design of bogie frame serves both fleets of vehicles and both motored and trailer bogies.

Braking System
The braking system composes three sub-systems.
- An electrically controlled pneumatic disc brake system blended with:
- An electric regenerative brake system using the traction motors as generators (a short term rated rheostatic resistor is used for transition duties and short periods of non receptivity).
- A spring applied air released parking brake system.

A Screw compressor on each unit supplies all pneumatic systems.

Doors
Plug sliding doors are fitted to both cab and passenger areas. All door leaves are bonded aluminium honeycomb skinned with aluminium sheet. They are operated by electric door motors with current sensing devices which control the door opening speed/force profile and also detect obstacles.

To minimise passenger stepping distances and the gaps between platform screen doors a critical review of track tolerances and assumptions inherent in the calculation was made and a separate kinematic gauge for stations was specified. The stepping distance even allowing for the extra factors arising from the use of plug sliding doors was maintained at the same as the existing MTRC service. Additionally the potential hazard of trapping children between the car body and the platform screen doors was minimised by the adoption of bi-parting door leaves which on opening move out from the body profile and fill the gap adjacent to the doors.

Cabs
AEL and LAL cabs are the same with the exception of the different controls for the Seat Back Passenger Information System. The cabs contain the Train Operators desk, Automatic Train Control equipment, Train Radio, all door control equipment, train traction and braking controls, and Control and Monitoring System controls. In addition, the cab front has the Passenger Emergency Detrainment door and fold-out/extendable ramp giving a tensioned Kevlar reinforced fabric walking surface (see figures 5 and 6 for details of the cab interior and detrainment ramp).

Control and Monitoring System
The Control and Monitoring System consists of a collection of intelligent electronic control equipment modules, distributed throughout the train, with the primary objective of data gathering, analysis and transmission of train data. The data is used to provide the Train Operator with adequate status information on the trains systems to enable speedy fault recovery, and also enable more detailed information to be made stored for subsequent use by the maintenance personnel.

Couplers
Multiple units consisting of 2 or 3 vehicles within the train consist are connected by auto couplers across which electrical and pneumatic control signals pass. Secondary coupling devices are fitted for extra security to all inter-unit automatic couplers. Cars within a multiple unit are connected by semi-permanent couplers across which power cables also are connected. Cab ends also have auto-couplers but they do not have pneumatic or electrical connections so as to prevent the possibility of an assisting train connecting onto a fault when attempting an emergency push-out.

Traction equipment

From the pantograph each traction power circuit comprises a high speed circuit breaker, line filter, traction invertor case and 4 x 3 phase traction motors. Each traction motor is rated at 245kw to meet certain interstation run times. This rating also allows the equipment to adequately satisfy the emergency pushout requirements. A pantograph supplies a maximum of two traction propulsion circuits.

Auxiliary Power Supply, Battery Charger and Emergency Invertor

Each unit within the train has two Auxiliary Power convertors, Battery Chargers, Batteries and Emergency Invertors to supply power for lighting, air conditioning, battery charging (and in turn all train control and passenger information systems), and emergency lighting and ventilation. Each Auxiliary Power convertor is rated at 130 KVA. The batteries are sized to supply emergency ventilation, lighting, communications and other essential train functions for upto 60 minutes in the event of an overhead power supply failure, necessitating a battery of 120 Ah.

Event Recorder

Each cab car is fitted with an Event Recorder which monitors and records operational data which would assist in an investigation of a reportable incident. The memory is capable of storing data from two round trips on the first in, first out principle, and is contained within a tamper proof assembly with restricted access.

EMC

To achieve compatibility with train control and protection and communication equipment, customer goods and appliances including cardiac pacemakers (there is currently no Hong Kong Legislation relating to radio frequency emissions for railway rolling stock) the standards employed are principally European with ENV50121 "Railway Applications - Electromagnetic Compatibility" providing most of the requirements. At the time of contract award the signalling contract was not finalised and it was necessary to include in the rolling stock specification preliminary limits for EMC in relation to signalling compatibility. These limits were revised when the precise nature of the signalling system was known.

Fire

To address the concerns of the Hong Kong Fire Services Department of running trains with soft furnishings and baggage in tunnels MTRC developed its own fire performance requirements based upon the most stringent international standards (BS 6853 for smoke and flammability, NF 16-101 for toxicity).

Noise

To achieve both a high quality passenger environment and to address Hong Kong Environmental Protection Legislation great attention had to be given to noise management throughout the specification.

To meet the specified requirements the car construction has included additional sound insulation beneath a wooden floor which itself is suspended on rubber blocks, plug sliding doors, double glazed windows, frame suspended motors with specially designed silencers on both the air inlet and outlets, and wheel face mounted brake discs to provide a degree of noise damping and also reduce wheel tread damage sometimes caused by wheel tread friction brakes. Similarly Automatic Train Operation can result in a high frequency of high brake rate demands and therefore wheel slide protection to UIC 541-05 in both service and emergency is specified to reduce the occurrence of wheel flats in the widely varying adhesion regimes experienced in tunnel and on the surface throughout the year. This is intended to contribute to noise reduction as well as maintenance cost for both train and track.

The specified noise level requirements with all auxiliary equipment running were:

	Cab Interior	Saloon Interior	Exterior @ 7.5m	Exterior @25m
Train Stationary	60dBA	60dBA	72dBA	-
Train Moving on surface 135km/h	73dBA	68dBA	-	83dBA
Train Moving in Tunnel 135km/h	82dBA	77dBA	-	-

10.2 Differences

Passenger Information System

Both train types incorporate illuminated route diagrams which show which station the train has arrived at and the side of door opening. In addition the AEL diagram shows the actual progress of the train down the line. The Public Address system incorporates a pre-recorded automatic Digital Voice Announcement System (DVAS) announcing station and door side opening and closing messages, triggered by the ATC equipment and also allows the operational Control Centre to make announcements directly to passengers via the train radio. Passenger Alarm Units located in every door vestibule allow the passenger to alert the Train Operation of an incident or distress and allow the Train Operator to permit two way communication. An emergency stop is not automatically initiated upon activation of the alarm but a train cannot leave a station without all alarms being reset.

In addition the AEL system offers a Seat Back TV System. LCD screens are located in most seat headrests and vestibules (as shown in Figure 12) and passengers can select from 9 channels giving various information on flight arrivals/departures, world and business news, weather, MTRC and Hong Kong traveller information. As distribution and collection of head phones is not practical within a 23 minute journey the system is configured to run without sound. Maintainability is enhanced by unit construction of the LCD screen and headrest allowing modular exchange at terminus stations.

Air Conditioning

Each car has two roof mounted air conditioning units capable of fully ventilating and air conditioning the passenger area. In addition the saloon areas can be ventilated in the event of OHL power failure by using the car battery and emergency inverter for a period of 60 minutes. The AEL has a total of 40kw thermal capacity per car and the TCL has 84kw per car. The cabs are also air conditioned.

Gangways
The cars are connected by means of inter-connecting flexible gangways. AEL cars have, in addition, sliding doors at each end of the car saloon. These doors are fire rated to give car to car protection of 30 minutes. AEL gangways are 700mm wide, TCL units are 1400mm wide affording passenger standing space.

Door and Window Configuration
All windows are direct bonded double glazed units with TCL windows incorporating emergency ventilation hopper units. The TCL door pitch (5 doors per car each 1400mm clear width) is maintained at 4500mm to align with the Platform Screen Doors. The AEL has two doors per bodyside each 1200mm clear width. The AEL doors are positioned at the second and fourth TCL door centres to allow the two train types to use each others station in degraded emergency operating conditions (Figures 7 and 9).

Interiors
The TCL interiors are a development of the existing MTRC vehicles with 48 seats per car and a crush laden capacity of 375. Figure 9 shows the interior layout of a TCL cab car which additionally incorporates two wheel chair positions at the expense of the length of two bench seats. Figure 10 shows the detail of TCL interior finish.

The AEL vehicles are soft furnished with fitted carpets and 64 seats per car with 4 baggage racks in each vestibule. The seats are fully cantilevered off the bodyside so giving stowage space for hand luggage underneath. Figure 7 shows the interior layout of an AEL cab car which additionally has two wheel chair positions with flip up seats at the expense of 4 conventional seats. Figure 11 shows the detail of the AEL interior finish.

Baggage Car and Automatic Handling Equipment
The AEL service carries passengers' check-in baggage in special 30 minute fire rated containers. To achieve the loading or unloading of the thirteen containers within the dwell time the TCL bodyshell was adopted for the AEL modified to incorporate the AEL gangway end (with secure inter-car doors), exterior livery, and no bodyside windows. Emergency egress is facilitated by means of a segregated gangway along one car side passing the containers mounted on the automatically controlled roller bed. The car is ballasted to compensate for this asymmetric loading. Baggage car door operation and subsequent container loading and unloading is controlled by the baggage handling supervisor after door enable signals are provided by the Train Operation (see Figure 8 for details of the baggage car interior and automatic container handling equipment).

11. CONCLUSIONS

It has been possible, by addressing the special market requirements of the airport passenger and the Hong Kong commuter along with the small fleet sizes for each service at the outset, to specify and procure trains for the two services with a minimum of equipment and maintenance differences. This allows both fleets to be operated and maintained using common facilities and practices and hence achieve economies of scale throughout the life of the trains. Furthermore concurrent design of the railway infrastructure has permitted optimization of the overall railway service in both design and degraded conditions.

Each of the two train types, though sharing the same cab and bodyshell profile, provide a very distinct brand image appropriate to its dedicated service both externally and internally - the AEL affording a business seating layout complete with seat back TVs and comprehensive baggage stowage areas necessary to encourage usage by air travel minded passengers whereas the TCL continues the existing Mass Transit Railway Corporation concept but incorporates the new Corporate identity, along with the improved noise and air-conditioning performance common to both train types.

The specified cab design has given the trains a strong identity on which to build a marketable image of the service and also has given the operators a much improved working environment. The use of large glazed areas has contributed to the very modern aspect of the cab front and has necessitated the development of a completely new form of detrainment ramp.

The introduction of the Seat-back TV system and the In Town Check-in Service has been an innovative step in firmly placing the AEL train in the same market segment as business class air travel and therefore reinforced the passenger experience of a seamless transmission from air travel to train.

Substantial achievement of the specified requirements, though providing the contractor with a great challenge, has resulted in the finished vehicles giving high levels of performance for both types of service.

12 ACKNOWLEDGEMENTS

The authors would like to thank the management of the MTR Corporation for their permission to publish this paper.

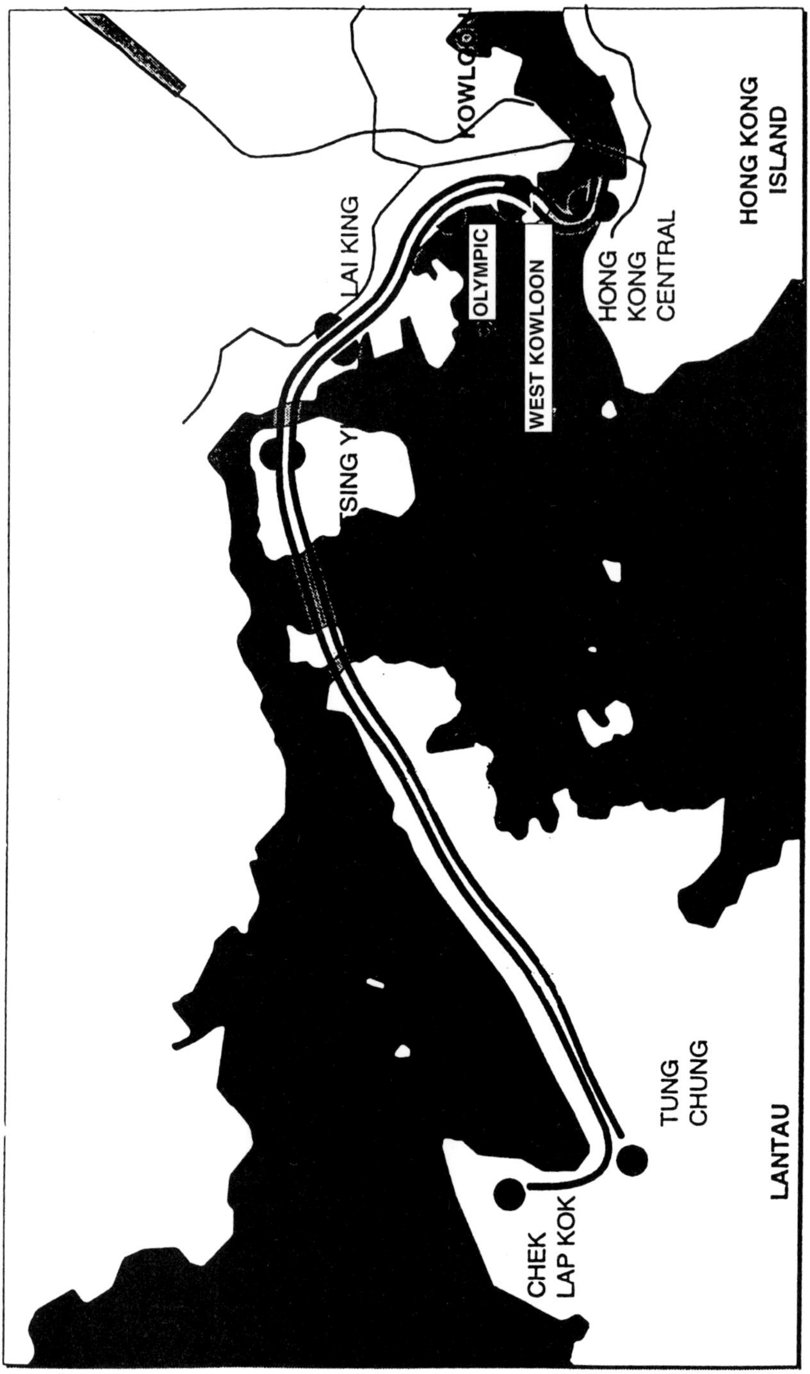

Figure 1 - Route Map of AEL and TCL

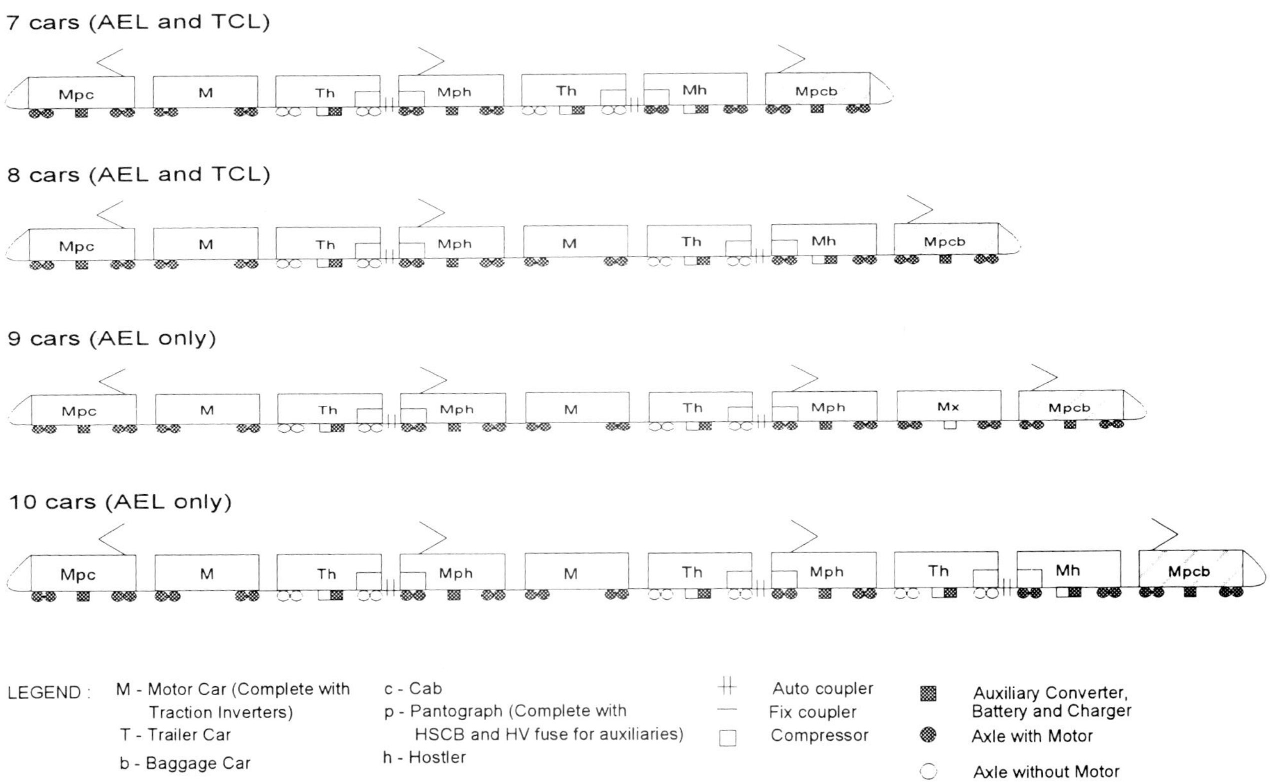

Note: On TCL trains, the Mpcb car is substituted by the Mpc car

Figure 2 - Train Consists of AEL / TCL

Fig. 3 AEL Train Service Pattern (Year 1997 - 2000)

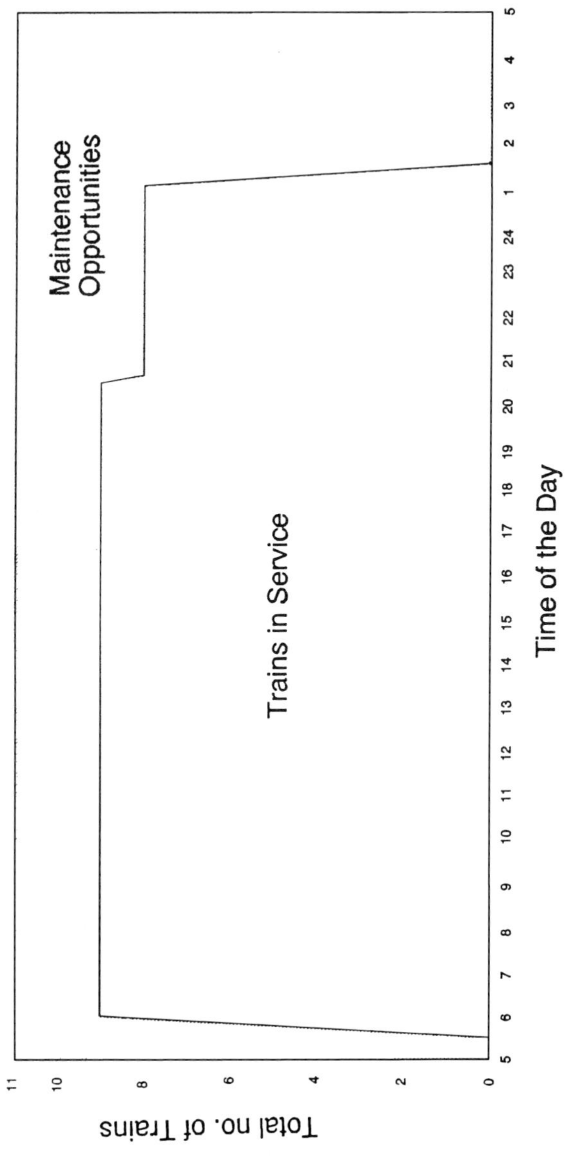

Fig. 4 LAL Train Service Pattern (Year 1997 - 2000)

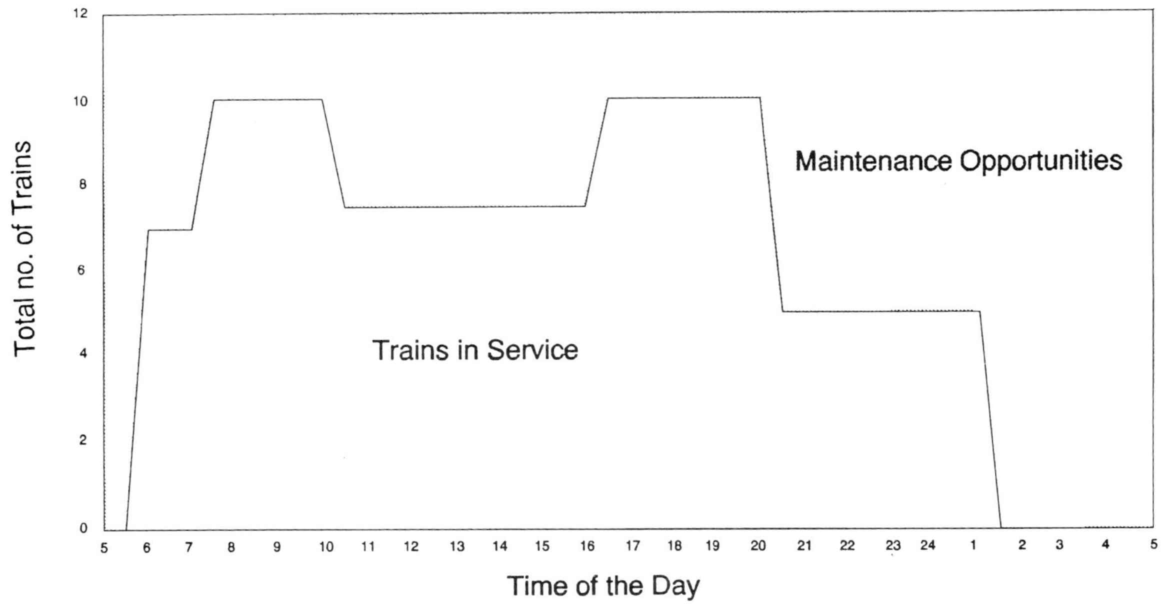

Figure 5 Photograph of Cab interior showing drivers desk and seat.

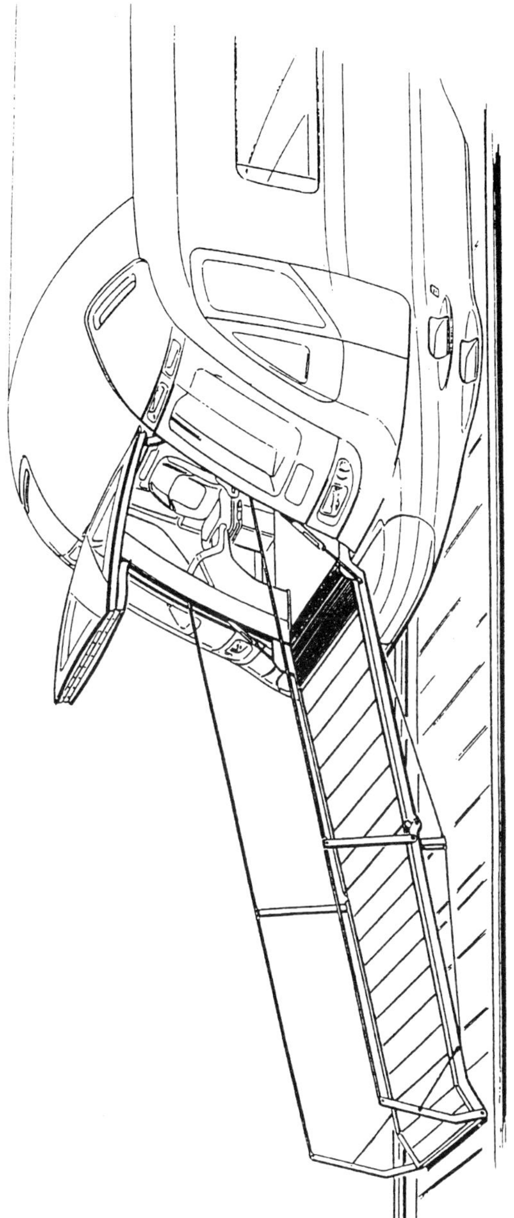

Figure 6 - Detrainment Ramp

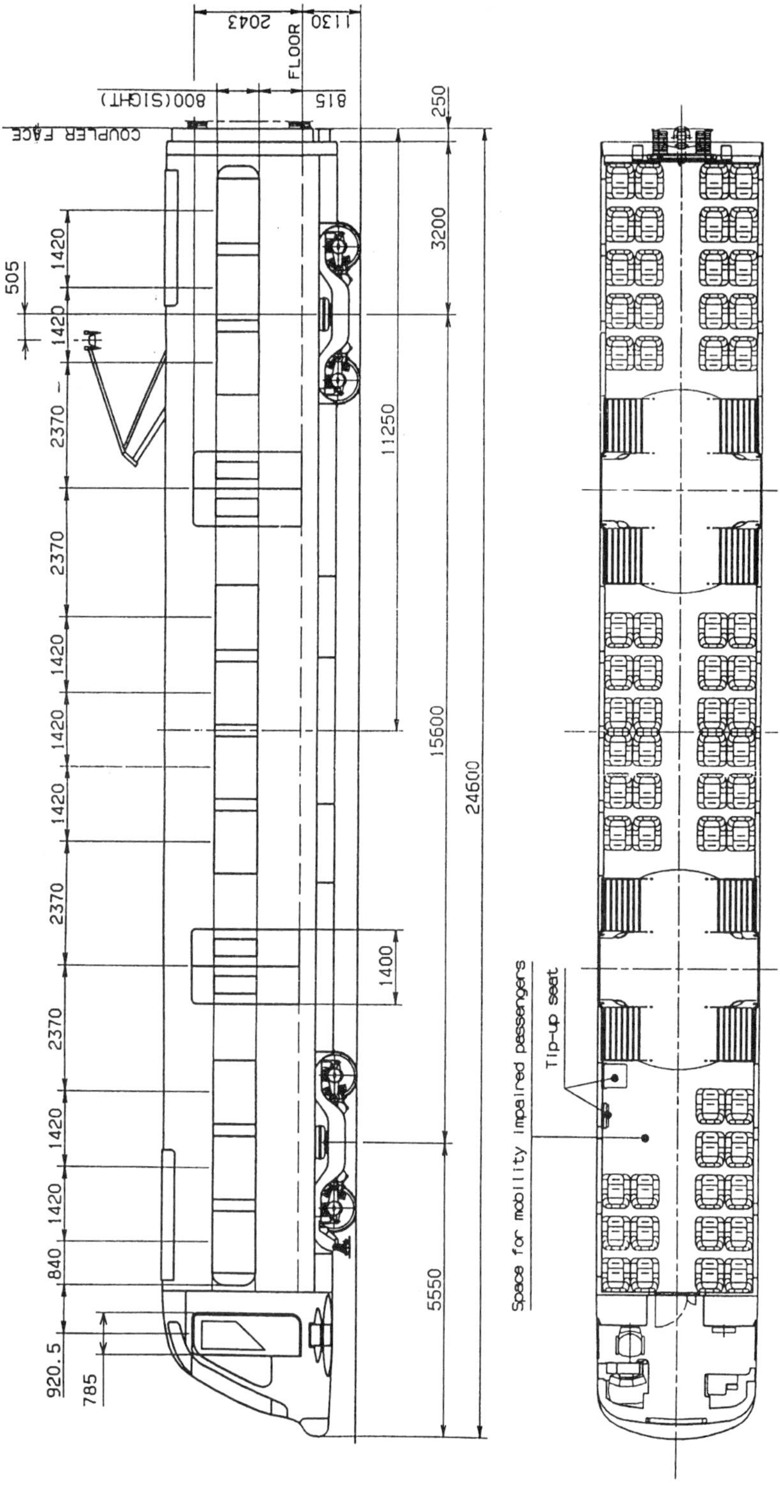

Figure 7 - AEL Cab car

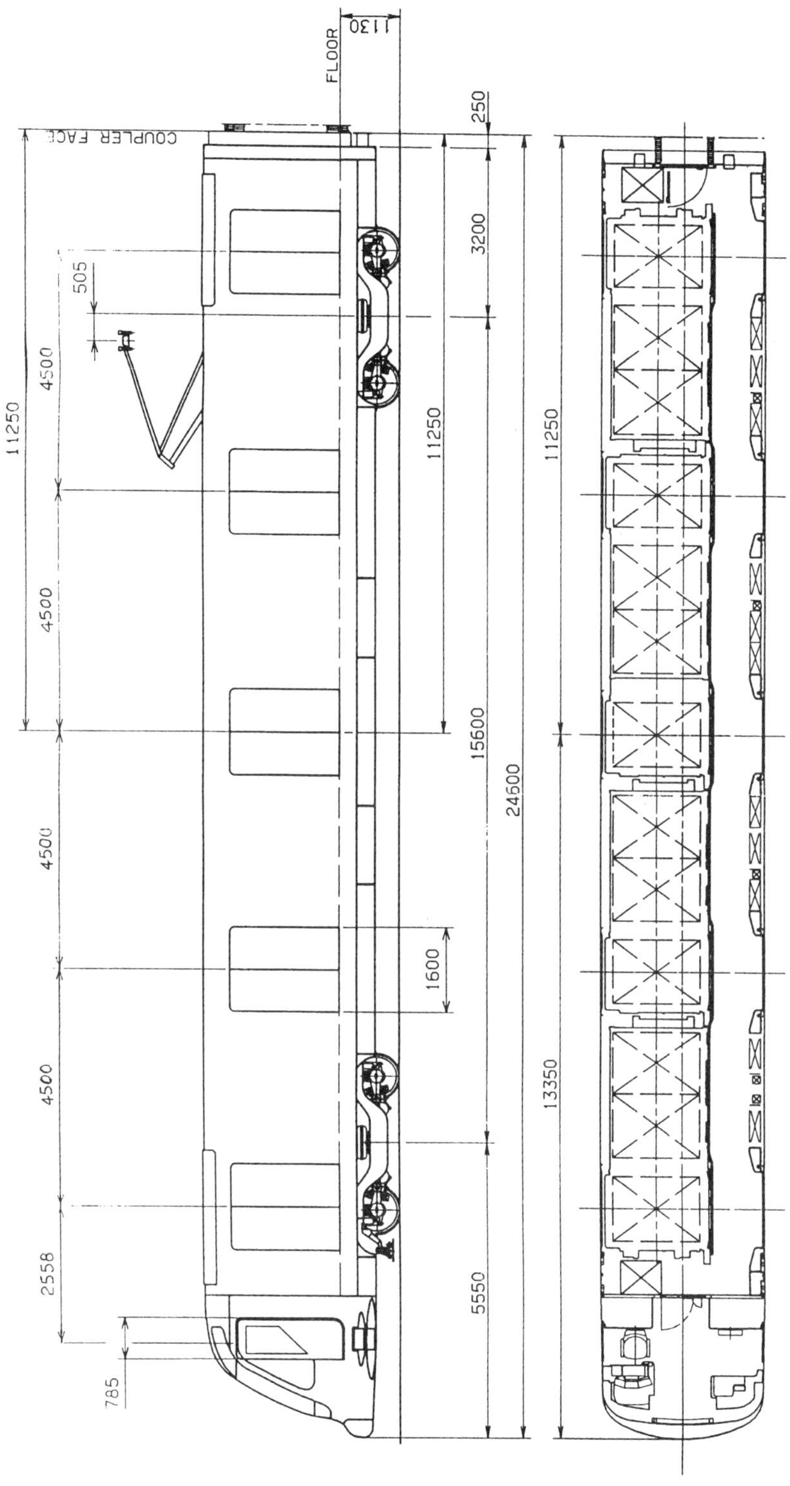

Figure 8 - AEL Baggage car

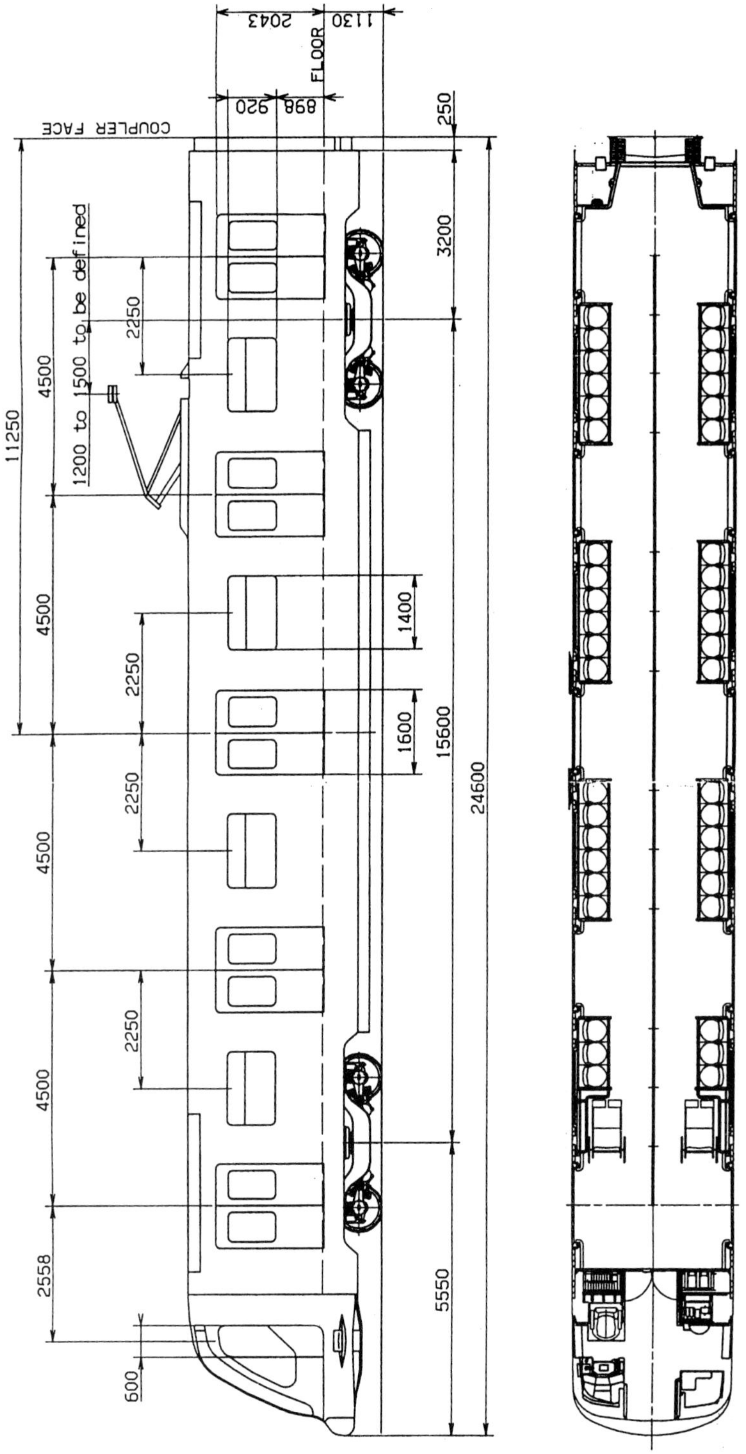

Figure 9 - TCL Cab car

Figure 10 Photograph of TCL interior showing vestibule, baggage racks, seats and gangway.

Figure 11 Photograph of AEL interior showing doorway, seating, and gangway.

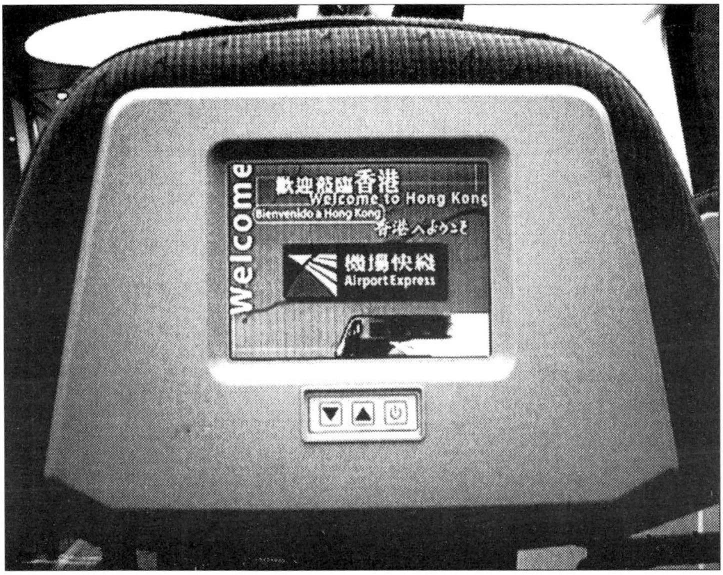

Figure 12 - Seat Back Television

Figure 13 Photograph of Vehicle Exterior.

C547/022/98

Getting the trains approved

A COOKSEY BSc, CEng, FIMechE, MICE, FIRSE and **R SHORT** CEng, MIEE, MIRSE
HM Railway Inspectorate, Health and Safety Executive, London, UK

For a new train to be introduced three approvals or acceptances are required. They are statutory approval under the 'Railways and Other Transport Systems (Approval of Works, Plant and Equipment) Regulations 1994, acceptance by the Infrastructure Controller of the train for operation over its infrastructure, and acceptance of the train operators safety case under the Railway (Safety Case) Regulations 1994. The paper describes the statutory approval process as it applies to trains, the process for acceptance of the train operators safety case and the relationship with the Infrastructure Controllers rolling stop acceptance.

1. THE APPROVAL OF THE TRAIN

The first part of the paper deals with the statutory powers which apply to the approval of the train. It sets out the procedures for the statutory approval and how this relates to the rolling stock acceptance process operated by the Infrastructure Controller. An outline is given of the standards which are applicable to new trains.

The History Of Statutory Approval

The first appointment of a government inspector to examine the safety of railways was by the then Board of Trade in 1840. At that time a number of privately financed railway companies were attempting to build and open new railways. Parliament had exercised control over the powers the railway companies needed to raise capital and to build the railway but had not thought it necessary to legislate to control the actual opening of the railway.

However, some of the new railway companies - no doubt looking for an early return on their investment or simply short of funds - were prepared to open before the works were properly finished. Inevitably, a number of serious accidents occurred and in response to the public concern, Parliament was forced to take action.

The concerns of 1840 were primarily that the civil engineering works of the new railway were either incomplete or sometimes inadequate. The construction of the trains was of a lesser concern. Therefore, the first legislation provided powers for the inspection and approval of the 'fixed' works. The requirement for the inspection and approval of new railway works was in order to protect the safety of those using the railway. The approval powers were developed over time by various pieces of railway legislation, to include signalling systems and electrification equipment as well as civil engineering works, and altered as well as completely new railways.

It was only with the most recent legislation that the general approval powers were extended to include trains. Some Acts of Parliament, which authorised the construction of individual railways, did include powers of approval for trains. Typically, these railways were underground railways or the newer 'Metro' systems. In addition, from the late 1980s the Inspectorate were being informally consulted by the British Railways Board on the design of many new trains.

However, the Inspectorate could only make recommendations and use its influence. Following the Clapham Junction collision, the lack of legal powers for the approval of trains was considered again and it was proposed that the necessary legal powers should be available.

The approval powers today

The present approval powers stem from the Transport and Works Act, 1992. It was under that Act that the 'Railways and Other Transport Systems (Approval of Works, Plant and Equipment Regulations) were made. These Regulations came into force on 5 April 1994. They brought up to date and replaced the earlier approval powers. The 1994 Regulations for the first time included general powers for the approval of trains although no mention of trains (or rolling stock) appears in the title of the Regulations. Trains are included within "Equipment".

The requirement for approval is a wide one. Regulation 4(1) requires that "approval shall be obtained before:

(a) any new works, plant or equipment; or

(b) any altered works, plant or equipment,

which are capable of materially affecting the safety operation of a relevant transport system, are first brought into use for the purposes of that system".

This approval can only be given when the works, plant or equipment is completely finished. However, the Regulations make allowance for works being brought into use before full approval is given. First, it may be necessary to bring works, plant or equipment into use in order to avoid interruption to the operation of an existing transport service. This is allowed for under Regulation 4(4)(a). It is more likely to be used for the introduction of a signalling system than for the use of a new train. Second, Regulation 4(4)(b) allowed with prior written consent for "(i) any testing or trials, providing that the testing or trials are concluded in accordance with the terms of such consent; or (ii) obtaining information to prove their satisfactory performance in connection with an application for approval." This is a provision which is of considerable importance to the process of obtaining approval of a new train.

Regulation 6 allows for an application by a manufacturer for type approval of a particular piece of plant or equipment providing it can be demonstrated that there is an operator who proposes to bring into use plant or equipment conforming with the type approved prototype.

The Regulations do not permit approval (except for approvals given for testing) to be given with conditions attached to the approval. Regulation 9 allows for the application for approval to include "relevant operational limitations". In other words, the approval is given for the use of the works, plant and equipment in the way specified in the application. If it is to be used in a different way, a further approval is required.

Approval under the Regulations is administered by HM Railway Inspectorate which is part of the Health and Safety Executive. Approval of trains under the Regulations is required in addition to any acceptance procedures undertaken by the Infrastructure Controller. The approval under the Regulations is concerned with safety aspects only.

Operation of the statutory approval process and its relationship with Infrastructure Controller's acceptance

It is the Inspectorate's intention that the legal approval process should operate with as 'light-a-touch' as is consistent with the use of its powers to safeguard those who travel and work on the railway. It is also the Inspectorate's intention that those seeking approval should not be subject to unnecessary burdens or delays.

The evidence supplied as part of a vehicle acceptance process to the Infrastructure Controller is likely to be satisfactory to the Inspectorate as well. Provided the rolling stock acceptance process operated by the Infrastructure Controller is transparent to the Inspectorate, the assessment made by the Infrastructure Controller is also likely to be accepted by the Inspectorate as part of the statutory approval process. That does not mean that the Inspectorate will not review both the evidence and the assessment made of it as part of its own processes.

The approval under the Regulations is concerned with the whole of the train and not just those aspects which involve the interface of the train with the infrastructure and with other trains using the infrastructure, which are considered through, for example, Railtrack's rolling stock acceptance process. With locomotives or freight stock there is little which will not be covered by the Infrastructure Controller's vehicle acceptance process. With a passenger vehicle a number of aspects, such as the internal layout and fittings, will fall outside the vehicle acceptance process. Therefore, the submission to the Inspectorate should also cover those aspects which are not covered by the Infrastructure Controller's acceptance.

The Regulations do not permit approval to be given with conditions other than 'operational limitations' specified in accordance with Regulation 9. The Approval cannot be given until the works, plant or equipment is complete. Under the 1994 Regulations, there is no provision for the giving of the 'provisional' approval, which was possible under the previous legislation. Instead, the Inspectorate now provides 'comfort letters' - a letter of non-objection - at various stages of a project's development.

In the case of a new train the Inspectorate would expect to see a series of submissions to which 'comfort letters' could be given for:

a. an outline of the concept of the proposed train (the 'business specification');

b. the development of the concept into an outline engineering design;

c. the development of the engineering design through 'mock-ups' and/or prototype construction.

At an early stage, the way the train is to be commissioned and tested, and how its safety is to be demonstrated should be discussed with the Inspectorate and the Infrastructure Controller. It is recommended that the arrangements proposed for these matters are formally submitted as a 'strategy document' for endorsement by the Inspectorate and the Infrastructure Controller before starting the more detailed work which will be required.

While it is helpful that as much testing as possible is done away from the operating railway, some of the final work will have to be done during normal service on the operating railway. This testing could, if not properly managed, endanger the existing railway and consent under Regulation 4(4)(b) is always required. The consent from the Inspectorate will not be given until a structured testing plan, which has the agreement of the Infrastructure Controller, has been provided.

It is not intended that approval under the Regulations is required for alterations that are undertaken as part of normal maintenance or the replacement of individual components or sub-systems if the replacement is of a proven design and does not change the behaviour in a way which will affect safety. Other alterations do require approval and the process is similar to that for a new train. The approval, however, is required for only the modification and does not require the remainder of the unmodified train to be re-approved. The Inspectorate is always prepared to give advice on whether approval under the Regulation is necessary.

The standards which apply

The Inspectorate will give approval under the Regulations provided it believes the proposal will provide an appropriate level of safety. From the early days of statutory approval the Inspectorate has produced written guidance on the standards it considers appropriate. Today that advice is contained in 'Railway Safety Principles and Guidance'. Part 1 gives the principles and outlines the factors which should be taken into account in considering how the principles should be met. Of the 33 principles, 10 are about trains. Further guidance on acceptable good practice is given in the Part 2 documents. There is a volume which deals specifically with trains. It is against these 'standards' which the Inspectorate will make its judgement.

The Infrastructure Controller will normally have industry standards which are likely to be more detailed than the Inspectorate's 'Railway Safety Principles and Guidance' and on which the vehicle acceptance by the Infrastructure Controller will normally be based. It is not part of the Inspectorate's role, as part of the approval process, to enforce the Infrastructure Controller's standards but they are normally accepted as an adequate way of demonstrating appropriate standards have been used.

The alteration of an existing train makes the choice of standards rather more complex than it is with a new train. The Inspectorate will judge proposals on their merit. The Inspectorate will expect that any alteration will give an improved safety level in so far as is reasonable practicable. It may not always be possible to meet fully the latest standards which would be applicable to a new train. The Inspectorate will also expect that consideration will be given to extending the scale of the work to give other improvements where that is reasonably practicable. However, there may be some situations when making small alterations, which are reasonably practicable, may not be sufficient. For example, minor modifications are unlikely to resolve the deficient crashworthiness of Mark I stock.

2 THE RAILWAY SAFETY CASE REGULATIONS

Since the earliest days of railways it has been a clearly established principle that the safety of a railway is the responsibility of the railway management. With the fragmentation and privatisation of British Rail the possibility arose that it might no longer be clear which railway management was responsible for controlling any given aspect of railway safety. This was of particular concern in relation to the separation of responsibility for infrastructure control and train operation, and the transformation of the former BR engineering departments into a multiplicity of private contractors.

In order to ensure that railway operators were aware of their safety responsibilities and that they would be capable of discharging them in a suitable manner, the Railways (Safety Case) Regulations 1994 require all operators of infrastructure, trains, or stations to prepare and have accepted a railway safety case which demonstrates their ability to operate with an acceptable level of safety.

The concept of a safety case, where someone seeking to demonstrate the safety of a system or piece of equipment by compiling a "case for safety" has proved to be very useful. The term "safety case" is now used in many contexts, including the documentation prepared in support of a submission for approval or acceptance of a new train. The wide use of the term can be confusing, and this paper aims to minimise the confusion by following the emerging convention of using the term "railway safety case" for safety cases prepare in compliance with the Railways (Safety Case) Regulations.

Scope of a Railway Safety Case

The Regulations require a Railway Safety Case to include, among other things, particulars of:

- technical specifications
- procedures or arrangements relating to operations and maintenance
- safety procedures for the design and procurement of premises and plant
- arrangements for ensuring the competence of employees as respects health and safety
- the organisation for carrying out safety policy
- arrangements for safety audit
- arrangements for managing work carried out by contractors

Railway Safety Cases must cover all these matters with respect to the design, operation and maintenance of trains, as well as other risk control measures.

The information to be included in the safety case may be categorised broadly as description of the operation, risk assessment, and description and justification of risk control measures. There is no set structure or format for a railway safety case. It is preferred that the operators concerned should structure their safety cases so as to optimise their usefulness as a safety management tool.

Whatever the structure of a railway safety case it should be possible to trace within it the elements of each of the systems used by the operator to ensure safety. The word "system" is being used in its broadest sense, and may include, for example, a system for controlling the movement of trains, a system for ensuring the competence of people engaged in safety critical work, or a system for managing crowd safety at stations. With regard to rolling stock, systems for ensuring that new trains comply with appropriate safety standards, and systems for maintaining trains in a safe condition, would be important components of a train operator's railway safety case.

The Assessment Process

Assessment of a safety case involves firstly a check to ensure that all of the information required by the Regulations has been supplied, and then a careful analysis to establish that the safety of the operation has been demonstrated. One way often used by HMRI for doing this is by seeking the answers to a series of questions designed to establish the safety of a particular function. Taking for example the safe maintenance of rolling stock, an assessor's list of questions might include the following:-

Is there a suitably structured, adequately resourced organisation to control maintenance?

- Are there adequate arrangements to ensure that people carrying out maintenance work re suitably competent?

- Are appropriate technical standards being applied?

- Are there effective procedures to ensure that scheduled maintenance is carried out?

- Is there an effective system for reporting and remedying defects?

- Are there adequate arrangements for monitoring and auditing performance?

The actual set of questions used will be adapted by the assessor to suit the circumstances of the operation concerned.

The Chain of Acceptance
The Safety Case Regulations provide for a chain of acceptances of safety cases as follows:-

An infrastructure controller must not allow trains or stations to operate until:-

(a) HSE has accepted the safety case;

(b) the train and station operators have agreed in writing to comply with any reasonable request made by the infrastructure controller to secure health and safety; and

(c) safety case have been accepted covering all the train and station operations.

A person shall not operate trains unless -

(a) a safety case has been prepared

(b) the safety case has been accepted

(i) by the relevant infrastructure controller at least 28 days before the operation commences; or

ii) where the train operator is the relevant infrastructure controller, by the HSE.

The chain of acceptance of safety cases ensures that there is consistency between the various operators on a given railway, with the infrastructure controller taking the lead in ensuring overall safety. It also ensures that safety cases are accepted by a body independent of the prepare and user of the safety case: no-one is allowed to accept their own safety case.

The 28 day period allows the HSE to scrutinise safety cases accepted by infrastructure controllers and to take action if any matters have not been correctly considered.

Role of the Infrastructure Controller

The chain of safety case acceptance, together with their responsibility for ensuring the safe movement of trains on their infrastructure, gives infrastructure controllers such as Railtrack or Eurotunnel a very important role in the acceptance of new rolling stock. They have to have accepted the train operator's procedures for the procurement of safe rolling stock and the safe operation and maintenance of that rolling stock. They also have to have procedures of their own for granting access to their infrastructure which will ensure that trains meet their safety standards, expressed as Railway Group Standards in the case of Railtrack, and are compatible with the infrastructure in respect of such matters as gauging, braking performance, train control systems, electromagnetic compatibility and so on. The Infrastructure Controller is also required to monitor and audit the compliance of the train operators with their railway safety cases.

The arrangements for vehicle acceptance, and for monitoring and audit, will form part of an infrastructure controller's railway safety case, and are thus subject to formal acceptance by HM Railway Inspectorate.

Duty to Conform

A railway operator, who has had a Railway Safety Case accepted, has a duty to conform with the procedures and arrangements described in it. Thus technical standards and installation and maintenance procedures included in a Railway Safety Case take on the status of legal duties, and failure to perform such duties would be an offence under the Health and Safety at Work Act. In this way the safety case regime has the effect of building many of the fundamental principles and practices of railway signal engineering into the legislative system.

Operating Procedures

When introducing new trains, an operator should review his procedures and arrangements relating such operational matters as train driving, despatching and platform duties, train coupling or dividing, emergency procedures, and any other arrangements which may be affected by the new design or changed facilities of the new train. Where changes to operating procedures are made in connection with the introduction of new trains, it may be necessary to obtain formal acceptance of the relevant changes to the railway safety case, if the changes are of such a nature as to materially affect the safety case.

C547/015/98

Building reliable, maintainable trains

C M PECKHAM BEng, AMIMechE and **H R SIMPSON** BEng, AMIMechE
Interfleet Technology, Derby, UK

SYNOPSIS

This paper discusses the availability, reliability and performance of new trains from the view of the operator. The paper illustrates some of the ways in which deliverables in these areas can be measured. It also examines ways of achieving the targets both technically and by the use of financial incentives. Based on Interfleet's experience supporting the Heathrow Express project, the paper discusses the importance of considering reliability and maintenance from the beginning and the potential penalties of failing to do so.

1. WHAT ARE THE TARGETS ?

1.1 Availability

The usual definition of availability used on the ex British Rail (BR) network is the proportion of a fleet of trains which is available to cover service diagrams. Availability is a function of fleet size, but is otherwise a measure of maintenance efficiency and reliability. At all times the actual availability (AA) must be greater than or equal to the availability required (AR).

$$AA = \frac{N-(S+U)}{N} \times 100\% \qquad AR = \frac{D}{N} \times 100\%$$

AA - Actual Availability % AR - Availability Required %
N - Number of units/vehicles/trains in fleet D - Number of diagrams
S - units/vehicles/trains stopped for scheduled maintenance - maintenance regime dependent
U - units/vehicles/ trains stopped for unscheduled maintenance - reliability dependent

Both AA and AR will vary throughout any given day. However, it is often most useful to quote the values at the time of peak AR.

Poor reliability and long maintenance downtimes result in low AA. The lower the AA, the larger the fleet size required to meet a given diagramming requirement. Examples of peak AR are shown in figure 1 for nine new or planned fleets of trains.

Figure 1 - Peak Availability Required (AR)

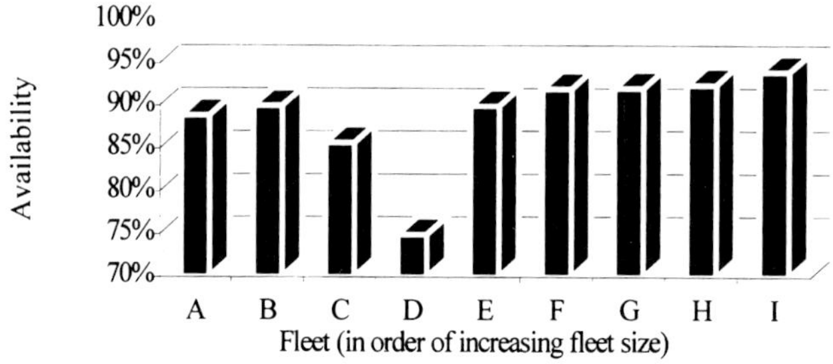

Fleet A	9 DMUs	**Fleet D**	18 'InterCity' trains	**Fleet G**	38 EMUs
Fleet B	10 'InterCity' trains	**Fleet E**	20 'InterCity' trains	**Fleet H**	40 'InterCity' trains
Fleet C	14 EMUs	**Fleet F**	25 Light rail vehicles	**Fleet I**	79 DMUs

When assessing the fleet size requirement, there is an element of balancing risk against cost. The balancing process will depend on the type of service provision, and any options that exist for drawing on a wider pool of trains. e.g.

- Cancel premier service - no alternative vehicles (e.g. Heathrow Express) ↑ **HIGH RISK**
- Cancel high frequency or small market service - no alternative vehicles
- Replace service with alternative vehicles (e.g. Mk2s instead of HST) ↓ **LOW RISK**

This explains the comparatively low AR for Heathrow Express (Fleet C) which is a premier service. It is also the only train certificated for operation on the route. Fleet D is a high speed service running in mainland Europe, again with limited options for alternative vehicles.

At the extreme, it may be possible to operate a service requiring 100% AA at certain times during the day, as long as :

- the maximum downtime for any maintenance activity will fit in between these periods.
- the geographical spread of operations enables this maintenance to be programmed.

Whilst this minimises both the capital costs of vehicle purchase and maintenance costs, there is no contingency in the event of train failure, collision damage and other unpredictable events. In the case of larger fleets (more diagrams) it is considerably easier to maintain high availability since the effect of each stopped unit or vehicle is comparatively less than for small fleets. The size of the 'Maintenance Unit' can also influence the ability to achieve higher availability. For example, a fixed formation effectively reduces fleet size compared with a fleet of interchangeable vehicles.

Care needs to be taken when comparing availability statistics, particularly if they are from outside the UK, since there are a number of alternative definitions in use. These include : -

- 'The ability to provide sufficient trains to meet all service requirements.' i.e. if there are 20 trains to meet 15 diagrams, and 15 trains are always provided, availability would be 100%. If the remaining 5 trains are stopped for maintenance then the UK definition would produce an availability of 75%! This definition can be particularly misleading since 100% 'availability' is achievable by having a lot of badly maintained, unreliable trains.

- $$\text{'Availability'} = 1 - \frac{\text{Sum of unplanned repair downtimes (hours) for fleet}}{\text{fleet size x 24 hours x days}}$$

  This measure does not truly reflect the needs of the operator as it does not account for scheduled maintenance. It also can be misleading because lots of small repairs will not heavily impact the calculated availability, where in reality this will cause havoc to service provision, e.g. if 20 trains each need 1 hour corrective maintenance in a 24 hour period :

  $$1 - \frac{20 \times 1}{24 \times 20} = 96\%$$

From the data quoted in figure 1, it is suggested actual availability (AA) targets should be aimed at 90% with a tolerance of about ± 5% depending on fleet size and on the type, and geographical spread of service provision.

1.2 Reliability

Reliability is a measure of delays (or the frequency of delaying incidents) attributable to technical failures. It has already been mentioned that poor reliability means poor availability. In addition it also leads to customer dissatisfaction, and penalties imposed by Railtrack (in the UK) and other bodies.

Like availability, reliability can be measured in a variety of ways and care is needed when comparing statistics. The most common measures are :

- *Miles (or km) per casualty.* There are numerous ways of defining casualty but the most commonly accepted criterion in the UK is a failure that causes a cancellation or delay of 5 or more minutes for ex BR routes, or 2 or more minutes for high density systems like London Underground.
- *Miles (or km) per potential casualty.* Similar to the above, but in some contractual agreements the definition of casualty is much more stringent (i.e. any in service failure that causes a cancellation or delay, or has in the past caused a cancellation or delay, or should it occur again foreseeably cause a cancellation or delay.)
- *Total minutes delay.* This gives a measure of the failures severity in terms of the operational consequences.
- *Mean time between failures.* This is another measure of the frequency of failures, but in this case it is useful where time is more important than mileage.

To monitor reliability most effectively requires a combination of measures. Miles (or km) per casualty gives the frequency of failures but not their impact on the service, and so should be used alongside total minutes delay.

When comparing reliability data it is important to understand exactly what defines a casualty or potential casualty and only compare like with like. The alternative definitions explain why some recent builds of train appear to have reliability an order of magnitude worse than others. Miles (km) per potential casualty may be fairer than miles (km) per casualty, since it removes operational inconsistencies from the measure. For example, a technically caused incident which causes a cancellation is treated with the same degree of seriousness as one which would have caused the same cancellation except for the fact there was a station fitter on the station to fix it before the scheduled departure time.

Reliability is dependent on design and build quality as well as maintenance. Components used in design should be chosen to only include items of proven reliability. Systems should be designed so that as far as is practical, no single point failure will prevent the train from reaching its final destination. Ease of maintenance is also an important factor in design for reliability.

Reliability of any new train will normally show an initial growth period, where component bedding in occurs, and teething troubles are addressed. Reliability targets are shown below for Heathrow Express. They are tough compared with most existing units but were considered achievable with modern technology and careful design. No one figure can be given as an industry standard since reliability that can be realistically expected will vary depending on the trains operation, distance, number of stops and at what point you measure the delay.

Time trains in service	Reliability km per casualty (5 minute delay)
3 months to 1 year	60 000 km
After 1 year	100 000 km

1.3 Performance

Performance is the ability to meet all the fare-paying customers' requirements. This includes providing comfortable seating, at the right temperature, in a clean environment. It is difficult to rank performance because to a degree it is subjective. Unlike reliability where it is possible to measure minutes delay, one person's perceptions of what is a quality environment may be entirely different from another's. However, some widely accepted performance criteria are : -

Criteria	Standard Required
Cleanliness	All passenger areas clean, and litter free.
Seats	Undamaged, and fit for purpose.
Lighting	Adequate lux level, no bulbs, tubes or covers missing.
Toilets	Clean, fully functioning, including handwashing and drying facilities.
Graffiti	Nothing offensive, or more than agreed limits.
Destination indicators	Legible, and operating correctly at all locations fitted.
Air conditioning	All passenger areas at a comfortable temperature in all ambient conditions. Good air quality.
Public Address	Clear, audible and at a comfortable volume.
Doors	Fully operational, open and close when required.
Notices etc.	Secure, in good condition, all information clearly visible.

2. How Can The Targets Be Achieved ?

2.1 Redundancy Of Equipment

System reliability can be improved by duplicating critical equipment, so that single point failures do not cause delay, cancellation or loss of customer facilities. The system becomes more tolerant to component failure. The disadvantages of introducing redundant equipment are an increase in the number of equipment failures, and an increase in the maintenance required. There is the also the risk of introducing hidden failures, that will not be discovered without additional failure finding maintenance.

The most efficient use of redundant equipment is therefore not complete duplication, but having a degree of system over-capacity, or an overlap between systems. The degree to which this is possible will depend on train formation. For example, in the two train formations shown below, option A has much greater potential for built-in redundancy than option B (Black wheels indicate motored axles) : -

Option A - Distributed Collection and Power Configuration

Option B - Loco Hauled Configuration

In the case of Heathrow Express, two identical auxiliary converters are provided on the unit to provide all 110V d.c. and 415V a.c. supplies required throughout the train. To provide the full load requirements both converters are required, but should one fail there is still sufficient capacity to supply all critical systems. Additionally, the use of a load shedding system means that air conditioning and other services can be operated on a cyclic pattern to maintain acceptable operating conditions.

2.2 Use Of Standard Parts

It is too large a risk in today's competitive environment to use equipment which does not have a tried and tested pedigree. The use of unique, and innovative technology may bring improvements, but is more likely to lead to reliability problems. The downside of this is it can lead to stagnation of design, but who should take the risk?

For Heathrow Express, the mechanical design was based on the class 446 and 447 EMUs built for Spanish Railways (RENFE) using well-tested components throughout the design. The bogies were modified to fit standard gauge track and the UK loading gauge.

2.3 Testing

Testing is an essential process in reducing residual risks implicit in any new train. This testing may comprise of component testing or complete train tests. In the case of Heathrow Express, the bogie design and traction equipment were tested thoroughly early on. The first bogies were fitted to a Mk 3 sleeper in April 1996 and tested on the Great Western Mainline (GWML) prior to the bogie frame going into full production. At the same time, dynamometer testing was used to prove the traction system.

Once the first units were substantially complete train testing was needed to ensure :

- compliance with all applicable standards
- electrical interference criteria were met
- the train performed to specification
- train reliability could be proven.

Velim, Czech Republic. The first train testing was carried out on units 1 and 2. Primarily the type tests verified the trains traction and braking performance, and electrical interference levels.

Wildenrath, Germany. During summer 1997 at Siemens' own track (a 6km loop), testing was carried out to prove the train's reliability and allow initial shakedown of components. Three units were run to a simulated service pattern, in six car formation. Lighting, doors, air conditioning and all other auxiliary systems were also operated under service conditions. As a requirement of the contract, 5000km fault-free running was required by each unit before acceptance by Heathrow Express. During the testing at Wildenrath approximately 7500km were completed by each unit, but unfortunately minor defects prevented the fault free target from being achieved.

Great Western Mainline. In September 1997, a safety case was achieved for testing on the GWML. Testing concentrated on Electrical Equipment and Control System proving tests (EE&CS). Initial tests were with one unit, but were developed to run a 'shadow service' operation immediately prior to 'FastTrain' passenger service introduction in January 1998.

2.4 The Importance Of Training And Documentation

The costs of providing excellent training and documentation can be recovered in the early stages of operation since unnecessary defects and failures can be avoided, and maintenance times reduced. Traditionally, training is left in the hands of the maintainer with little supplier support. Due to more pressing needs it is often the lowest priority of vehicle provision during the commissioning period.

Where new trains are being introduced into an existing depot, the main priority is to keep existing trains running and additional effort is needed to ensure maximum benefits are gained from training. Heathrow Express did not have this problem since the set-up was "new trains, new depot".

The manufacturer will usually be contracted to supply maintenance and spares documentation for the trains they are providing. The potential pitfalls here are :

- The supplier is not in the best position to fully understand the maintainers' facilities, environment or organisation, consequently the maintainer may end up re-writing the documentation into a more operationally useful form.
- Documents and spares information are often supplied late and may be incomplete.
- Documentation and drawings are not fully maintained e.g. modifications during build are not reflected in the accompanying documentation.

In the case of Heathrow Express provision of documentation was sub-contracted out by Siemens, which relieved some of the pressure on them. Even so, with this arrangement the core information still needs to be provided by the manufacturer in good time. Good communication between maintainer and author is vital to ensure the documentation meets the needs of the staff who will be using them on a daily basis.

2.5 The Value Of Life Cycle Data And Defect Monitoring

Even with the best possible maintenance information it will be difficult to get it right first time. It is necessary to review it regularly throughout the trains life in the light of service experience. Techniques such as Reliability Centred Maintenance (RCM) and Maintenance Regime Analysis (MRA) can be used to optimise the maintenance plan to prevent :

- Over Maintenance. e.g. examining something every day when once a month would capture any imminent failures costs unnecessary money.

- Inappropriate Maintenance - doing the wrong thing to address the problem e.g. examining and cleaning a component monthly, when renewing it annually would be more effective. Re-focusing the maintenance may reduce costs as well as improve reliability.

- Under Maintenance. i.e. not doing something as often as necessary, or not doing it at all. This will lead to poor reliability, safety risks, and probably in the long term costly corrective repairs!

To correctly assess which are the correct maintenance requirements it is desirable to have a detailed source of data to analyse. The supplier should provide life cycle data for components where this exists. During train testing and then service running detailed records of component condition, examination results, condition on overhaul etc. should be retained. These can then be used to feed into maintenance reviews on an ongoing basis.

2.6 Spares Supply And Warranty

A supply of the correct spare parts is needed at the correct time to ensure availability targets are met. Where parts are not available the practice of robbing parts from other vehicles will usually take place. This is difficult to control and leads to duplication of costs (the part must be fitted twice), and should therefore be avoided as far as possible.

Availability can be heavily impacted if insufficient spares are available. This can be a particular problem during services introduction and the early stages of operation. The reason for this is difficulties in setting up spares supply and warranty contracts and also the influence of unknowns in component failure rates causing short term material shortages.

3. WHAT ARE THE INCENTIVES FOR ACHIEVING THE TARGETS

3.1 Boundaries

Fundamentally, where the boundaries lie does not alter the issues of how to build and maintain trains for reliability; it just changes the incentives applying to the different parties. The diagrams below are a simplified model of the boundaries present in the most common structures.

i)

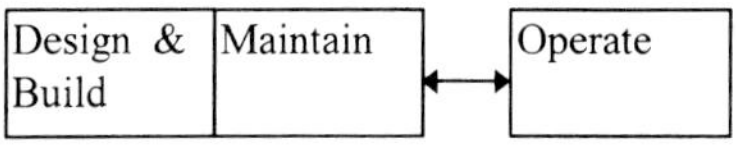

e.g. Heathrow Express

ii)

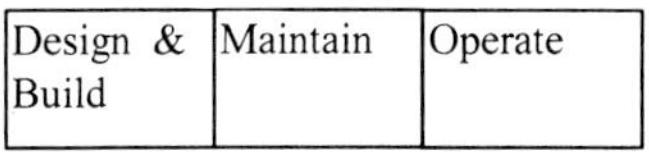

e.g. 1970s British Railways (BR)

iii)

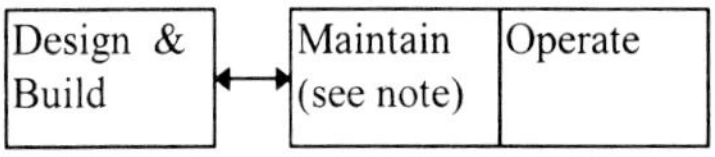

e.g. English Welsh & Scottish Railways

Note : This is similar to most ex BR routes with the exception that trains are owned by leasing companies (ROSCOs). Light maintenance is the responsibility of the train operating companies (TOCs), but heavy maintenance responsibility lies with the ROSCO.

iv)

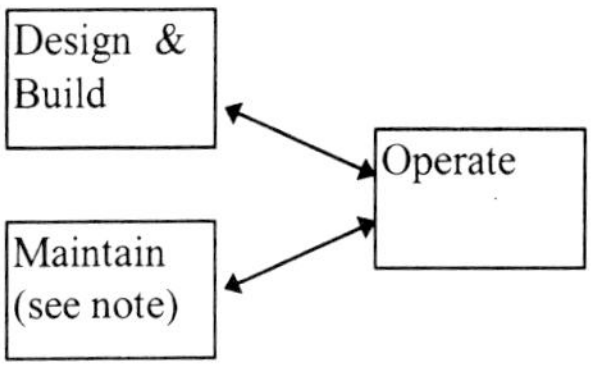

e.g. Thameslink

Note : Thameslink subcontract their maintenance to a third party.

Incentives need to be present between all parties to ensure the targets are met. The nature of the incentives will depend on the organisational structure between the parties, but may be : -

- Direct between parts of the same company or group of companies
- Enforced where they are contractual agreements between companies
- Inferred where they are common business sense between companies

For example in option (i) above : -

- There are direct maintenance incentives between the manufacturing part of the company and the maintenance part. If the trains are difficult to maintain, this directly costs the company money because of increased maintenance costs, and expensive modification programmes.

- There are inforced incentives between the maintainer and the operator to ensure the operator gets the right number of trains each day, in the required condition to run a reliable service. Enforced incentives usually take the form of some kind of penalty and bonus scheme. Excessively punitive penalty provision in tender invitations will result in two things, no bids or high prices. The level of penalty must be reasonable.

- In addition there are inferred incentives, if the trains are unreliable the manufacturer will find it difficult to sell any more!

3.2 Enforced Availability Incentives

Penalties can be imposed by the rail regulator if the operator consistently does not provide the service agreed in the franchise agreement. In addition, the losses to the operator can be estimated for the cancellation of any given service. In recent work relating to the purchase of long distance high speed trains, the cost to the operator for the cancellation of one train was put at £6000 in lost revenue. This sort of penalty will not be tolerated by the operator, but can be passed back to the manufacturer through the various penalty schemes written into the contract.

3.3 Enforced Reliability Incentives

In the UK, Railtrack will charge the operator for their delays to compensate other operators for knock on delays. The cost per minutes delay charged will depend on location and time of day. For example a failure in the middle of the rush hour at the entrance to Waterloo is going to be very expensive!

Typically the average cost per minute delay will be up to £40. Again the operator will wish to pass such penalties back up the ladder towards the manufacturer. This is difficult for the manufacturer to grasp because they are many parameters outside their control that effect the severity of the delays resulting from a given failure. Also because of the differing measurements of delay (Railtrack impact minutes, passengers charter, actual knock on delays) there is no true reflection of compensating the actual costs incurred.

Where the operator is not the maintainer, benchmarking reliability against existing trains can produce a fair penalty and bonus scheme. For example, if current fleet reliability is 200 minutes delay per 4 week period, and there is a requirement to increase reliability for the new trains to say 150 minutes per 4 week period. The bonus scheme could take the form :

- Maintainer pays operator £40 per minutes delay over the 150 minute target
- Operator pays maintainer £40 per minutes delay under the 150 minute target

This is fine for design, build and maintain contracts, but what about when the operator is maintaining? In theory, the same award penalty scheme can be operated between operator/maintainer and the manufacturer, but then there is always the argument that the reason their trains are no good is the maintenance is not being carried out correctly.

3.4 Enforced Performance Incentives

Performance can be monitored, and penalties awarded for failing to reach agreed targets. In the UK, this may be an agreement between the Office of Passenger Rail Franchising or a Passenger Transport Executive and the operator, where funding has been provided. Alternatively, where maintenance is contracted out penalties may be payable by the maintainer to the operator.

For a successful penalty / bonus scheme a system including the following elements need to carefully agreed : -

- Sensible targets for each criteria. (e.g. lighting, heating, public address etc.)
- Suitable method of inspection and auditing against the targets.
- Mechanism for converting performance against the targets into pounds penalty or bonus.

4. The way Forward

The most important lesson learned during the Heathrow Express project is there are many issues that effect reliability, availability and performance and these issues first become apparent long before service introduction. These are not often easily solved, but the earlier teamworking begins the more chance there is of achieving the targets when the trains start to run.

It is suggested that from the beginning of a train supply contract that manufacturers, maintainers and operators communicate effectively with each other, and work as a team. One possible solution would be to set up a joint team with appropriate funding to develop maintenance and reliability control systems. The team would be set up as soon as the contract is let, and remain in place at least through the initial years of service. It would comprise of representatives from the manufacturer, maintainer, operator, leasing companies (if applicable) and independent consultants as required. The team would quickly develop its own identity set apart from the individual companies.

5. Conclusion

The key targets for the operator are : -

- High availability
- Good reliability
- Good performance

These targets can only be achieved by : -

- Getting the technical issues right, i.e. design, testing, training, documentation, spares supply etc.
- Ensuring the right incentives are in place between the parties to ensure they work together to get it right.
- Successful teamwork.

6. Acknowledgements

References to Heathrow Express and Siemens are made by kind permission of the Engineering Director (Brian Raven) and the General Manager, STSL (David Wilson), respectively.

C547/021/98

New trains for old – heavy mid-life refurbishment of KCRC's EMU fleet in Hong Kong

M W Q YEOMAN BSc, CEng, MIMechE, FIEE, MHKIE
East Rail Rolling Stock, Kowloon Canton Railway Corporation, Hong Kong

SUMMARY

The East Rail Division of the Kowloon Canton Railway Corporation (KCRC) is in the business of moving people. Although only 35 km in length, some 720,000 people are now carried daily using a fleet of 117 3-car 25 kv EMUs, supplied by Metro-Cammell (UK) over a period of 9 years from 1982, and run in 12-car formations. Morning peak availability equates to 27 trains, or 92.3% of the fleet. At the peak between 07:45 and 09:30 trains run at 3 min headways, each carrying some 4,000 people.

With passenger growth rates of some 5% per annum, it was obvious in the early 1990s that an Automatic Train Protection (ATP) System would need to be installed to increase the Railway's capacity, together with giving improved levels of safety. However, during the planning stages of ATP, it became evident that station dwell times were becoming critical, and in order to derive full benefit from ATP, some means of reducing dwell times would have to be designed.

Accordingly a scheme was developed to increase the number of doors on each vehicle side from three to five. The opportunity would also be taken to refurbish the interiors, reform the trains into 12-car units, and upgrade selected equipment. The objective was to provide a train well-suited to even heavier usage, and capable of being safe and reliable for at least another 15 years.

An agreement was established with the original car builders, GEC ALSTHOM Metro-Cammell, to carry out the main part of the works, under the auspices of an "open book" approach ensuring full visibility of costs. A contract was signed in October 1995, and the first 12-car train re-entered passenger service in Feb 1997, having taken 7 months to refurbish. The contract requires that the entire KCRC EMU fleet will be refurbished by the end of 1999.

At the time of writing, seven trains are in service, and performing very well. Customer feedback has been generally very positive.

KCR East Rail is in the process of revitalizing its fleet to meet the demands of the new millennium, with an award winning product at less than 30% of the cost of a new train.

1. INTRODUCTION

In 1910, a single track railway with seven stations was opened, connecting urban Kowloon with the China border at Lo Wu. The British Section of the Kowloon Canton Railway was thus born and by 1920 was carrying just over a million passengers annually.

In 1983, following extensive modernization, double tracking and electrification, the infrequent diesel service gave way to regular interval services run by electric multiple unit trains every 4 minutes (peak) and half this frequency off peak. In this first full year of electrification, patronage had risen to some 23 million. Today the railway carries a staggering 720,000 daily, or 260 million annually! This figure is set to continue rising, as new towns adjacent to the railway are enlarged. Current forecasts predict a growth rate of over 5% per annum (see Fig. 1).

Fig. 1 : ER Patronage

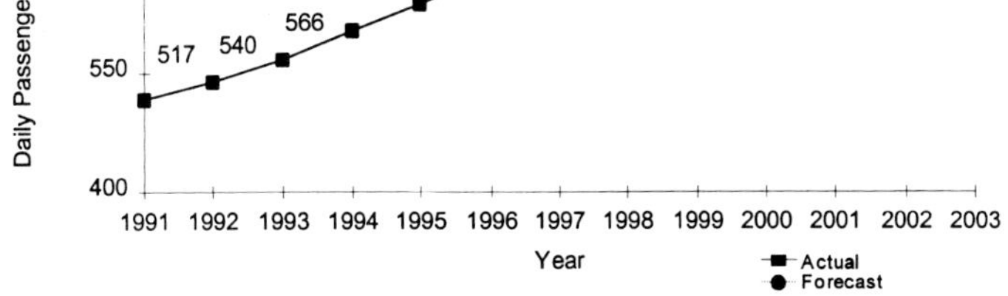

The original EMU fleet of 61 x 3 car units purchased from Metro-Cammell for the opening in 1983 has now been virtually doubled, giving a total fleet size of 351 vehicles. All trains are now formed by 12 cars, running at 3 minute intervals in the peak, and 6 minutes off-peak. Crush loaded, each 12-car train carries some 4,000 passengers. Currently to meet operational requirements 92% of the fleet (324 vehicles) must be available for the morning peak and slightly less (312 vehicles) for the evening peak.

Fleet reliability has risen to the point where minimal remarshalling of the 12-car sets has to be done because of train defects (see Fig. 2). Indeed, in 1996 when the driving cabs were refurbished, a decision was taken to refurbish only 58 cabs : 29 North and 29 South; the remainder being taken out of service. This was a leap of faith at the time, but has since proved fully justified.

Fig. 2 : EMU Reliability

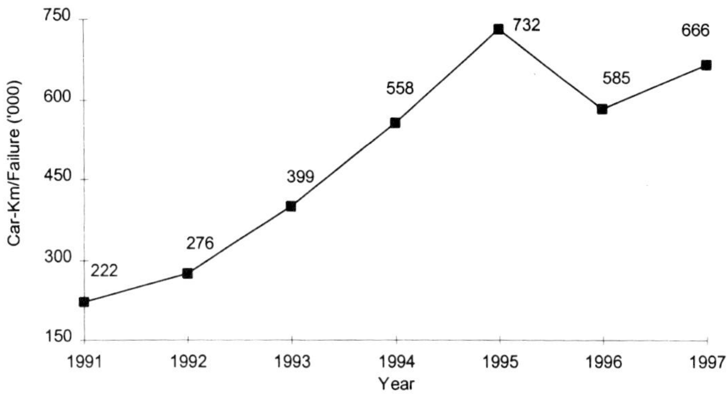

2. DRIVING FORCES FOR CHANGE

Because of increasing demand, KCR is in the business of forecasting growth, and planning how this growth can be economically accommodated. Of major importance is the installation of Automatic Train Protection (ATP) equipment, due for commissioning in mid 1998, which will enable KCR to increase peak services from 19 to 24 trains per hour.

However, in the midst of the ATP project planning process it became evident that with headways reducing from 3 minutes to 2½ minutes, station dwell time becomes critical. As this is determined by the rate at which people board and alight, so it became evident that, if we wanted to really benefit from ATP, then the number of doors on each EMU car should be increased.

Related also to our burgeoning passenger numbers was the internal condition of the cars. We calculated that, since entering service, each car had carried some 4.5 million people! Even with a rigourous cleaning regime, the cars were looking tired internally, and not at all in keeping with the vibrant, bright and clean appearance of our redeveloped stations. They certainly did not fit the image that KCR wanted and was driving to become, or that which our customers expected as we entered the 21st century.

As mentioned previously, only 58 of our 200 cabs had been refurbished. A further benefit to increase capacity would be to remove these redundant 142 cabs, which together with improvements in seating arrangements could provide us with some 12% more capacity.

The final driving force was that of EMU reliability. Various customer surveys had demonstrated to us that customers hold reliability as their main reason for travelling by KCR train. This reliability had risen dramatically from 1991 to 1995, and although still very high by world standards, it was evident that problems related mainly to equipment aging were on the increase.

The purchase of new vehicles could not be justified, due to the average age of the fleet being only 8 years. The opportunity to undertake a major refurbishment of the fleet, driven by the desire to increase the number of doors, together with increasing car capacity, therefore presented itself. Further benefits accruing from improved safety, reliability and reduced maintenance costs, assisted in persuading the KCRC Board that here was a project worthy of their support.

3. PARTNERSHIP

At an early stage in the project it was evident that a traditional tendering exercise was unlikely to enable KCRC to obtain best value for money. Indeed, because the Intellectual Property Rights (IPR) of the vehicle design rested with the car builder, Metro-Cammell, and they had no desire to hand these over to another, it was clear that the technical risk involved was likely to prevent other contractors from expressing interest. This was certainly true in relation to cutting four apertures in each car's aluminium body in order to fit additional doors, and our informal attempts to interest other Rolling Stock Contractors with refurbishment experience were fruitless.

There was also the issue of time. As ATP was likely to be installed in 1998, and passenger projections necessitated increasing availability from 92% to 96% from year 2000, a specific window of opportunity thus presented itself. The refurbishment of the fleet had to be finished by 1999.

So it was that in September 1994 the KCRC Board gave the green light for a "Partnership" approach with GEC ALSTHOM Railway Maintenance Services Ltd (GARMS) to be negotiated. GARMS is a sister company to GEC ALSTHOM Metro-Cammell and therefore had full access to the original design information and relevant design personnel.

The objective was to jointly develop :

- A technical specification for the works
- A timeframe for the works
- A form of contract acceptable to both parties
- A method of ensuring visibility of costs, together with a method of auditing such costs

A vital element in the partnership was the project staffing arrangements. GARMS had supplied a staffing plan showing engineering staff to be based in Hong Kong and the UK, as part of their original proposals. It was clear to both parties that the success of the job would be determined by the engineering work carried out in the 12 months immediately after Contract award, and therefore a GARMS Project team of the highest calibre would need to be assembled in Hong Kong. This was done, and a team of some 7 engineers headed by the GARMS Project Manager took up residence in KCRC Headquarters in January 1996.

The KCRC internal rolling stock design team was complemented by 6 engineering consultants from Halcrow Transmark. Project Management was initially provided by the KCRC Rolling Stock Design Manager, but was later transferred to a member of the Transmark team as the workload increased.

4. THE TECHNICAL SPECIFICATION

In addition to KCR's specific requirements, it was felt that the refurbishment would give an opportunity to generally ensure the vehicles were fully fit to meet the demands placed upon them until their planned life of 30 years was reached.

Accordingly, a team of four GARMS engineers came to Hong Kong in Sep 1994, to conduct a Fleet Audit, encompassing all equipment fitted to the EMUs. Consideration was given to the following questions during the audit :

- Will the equipment last the vehicle life?
- Should it be redesigned or replaced with an improved version to give enhanced performance/better reliability/less maintenance?
- Should it's existing maintenance regime be changed to improve life/reliability?

The audit lasted 1 month, and was concluded with an audit report, which was discussed in open forum over two days with the auditors and relevant KCR Engineers. Most of the recommendations were adopted, and a number included in the refurbishment project.

The high level technical specification was split into KCR works and GARMS works. The philosophy for splitting the works was that GARMS would be asked to carry out items which necessitated a specialist knowledge of the original EMU design, for example the additional doors, the thyristor traction equipment and the interior.

Items which were not critical, or had no particular programming difficulties, or where competitive tender would secure the best deal for KCRC, were tendered out in the normal way. Such items included the air conditioning system and wheelslide systems.

The agreed technical specification was as follows :

4.1 KCRC Works

- Refurbishment of Air Conditioning Equipment
- Refurbishment of Traction Motor Armatures
- Fitting of Hopper Windows
- Renewal of High Tension Cable
- Replacement of Traction Motor Gearcase with a GRP Variant
- Replacement of bogie brake pipework fittings with stainless steel alternatives
- Replacement of Wheelslide System
- Replacement of Battery.

4.2 GARMS Works

- Design and installation of 4 extra doorways to each car
- Design, procurement and installation of electrically operated doorgear
- Removal of redundant cabs
- Design and installation of new, modern, cab front-ends
- Fitting of linear fans to improve ventilation
- Removal of tapchanger equipment, replaced by GEC thyristor traction equipment
- Design, procurement and installation of emergency detrainment devices
- Renewal of motor coach bolster beams
- Design and installation of a new interior.

The final item is worthy of special mention. The internal design of any passenger carrying rail vehicle is of great importance, both to the passengers themselves, and to the Railway operators and maintainers. It is often the cause of much heated debate, as colour schemes and layouts are of course somewhat subjective. Because of the tight timescale, and knowing that this item was on the critical path, a small group of operators and maintainers, together with a designer from Jones Garrard, got together for several days of continuous meeting in Autumn 1995. Their remit was simple - to document an agreed performance specification for the refurbished EMU interior, and they had 10 days in which to do it! In the event they produced a seven page specification, which was owned by them all and has stood the test of time. Indeed, so impressive is the final product that Jones Garrard won a "Design Effectiveness Award" (presented by the UK's Design Business Association) in 1997 for our jointly envisioned design. Drawings and a full size mock-up were produced, and representatives from interested parties (including passengers themselves) invited to make comments, most of which were eventually incorporated into the finished product.

From developing the performance specification for the interior in Oct 1995, drawings and mock-ups were produced by Jan 1996, and the whole design frozen and essentially complete by Mar 1996. We believe this process to be world class, and a further endorsement of the effectiveness of the partnership approach adopted.

A significant additional item by necessity was added to the technical specification after the signing of the Contract in Oct 1995. This was to rewind the main transformers of the earliest cars in the fleet (60 motor coaches), and add molecular sieves to all transformers. Although problems were being experienced with Transformers during the development of the Project, the analysis work was still in its infancy, and the extent of the problem was not known during project authorization.

Suffice to say that significant degradation of the winding insulation, proved by furfural analysis of oil samples, signalled clearly that the windings would be unlikely to last the life of the EMU, and remedial action was therefore necessary. A new regime on transformer oil testing and maintenance has been developed, and will be the subject of a future paper.

That the project could absorb this substantial addition (a cost of some £4 million), with only a marginal exceedance of the project limit, again is a positive endorsement of the partnership approach.

GEC Transformers were awarded the contract, as a sub-contractor to GARMS, and the first new winding was fitted in May 1997 without causing any delay to the main programme.

5. SAFETY

At any early stage in the Project consideration was given to the management of safety. A U.K. based consultant was selected to assist in the process, and it was concluded that the preferred approach was the production of a Safety Case to cover the significant changes to the EMU caused by the refurbishment.

At this time, Hong Kong's Railway Inspectorate had a "dotted line" linkage with the U.K. Inspectorate, who were familiar and comfortable with the Safety Case approach. Furthermore, although there was no legal requirement in Hong Kong for such an approach, it was felt that the process of compiling a Safety Case was rigorous, proactive and professional.

The scope of the Case encompassed all operational safety issues from contract award through to eventual fleet operation. However, construction and commissioning safety issues arising from the project were specifically excluded from the Safety Case. These aspects were instead subject to the requirements of a separate safety plan, (drafted by GARMS and commented on by KCRC before issue), to ensure that such risks were reduced to as low as reasonably practicable.

The nature of the safety assessment undertaken for the Safety Case was incremental rather than absolute. Thus the assessment started with the premise that the current EMUs provide an acceptable risk to passengers, staff and members of the public. As a consequence, the objective of the Case for the EMU refurbishment was to demonstrate that levels of safety are no worse, and preferable better, for the refurbished stock when compared with the existing trains.

GARMS played an integral part in the production of the Case, as very often they had to provide evidence proving safety of particular equipment or processes. They were fully involved in the production of the Case, and were present at most meetings. This ensured joint ownership of the process, which we believe worked very well, and has been used subsequently on other projects. The Case is reviewed when necessary to ensure the assumptions are still valid.

I do not intend to describe the process by which the Safety Case was produced, as this has been covered previously (see References).

6. THE CONTRACT

As speed was essential, the Contract was developed in two parts. The first was termed the "Fixed" part, and referred to those elements of the works where the specification was clear, and as such a fixed price could be given by GARMS. The fixed elements were :-

- The replacement of the tapchangers with thyristor traction equipment
- The addition of 4 additional doorways per car
- The decommissioning of the redundant cabs.

GARMS presented their prices for these works in an open book manner, fully broken down; negotiations followed through questions and answers, and eventually a price was fixed, with GEC taking the technical risk in the normal way.

The second part of the Contract was defined as the "Variable" part, and referred to those elements of the works where the specification was in a formative stage, and thus insufficient detail was known to produce a fixed price.

GARMS acted essentially as a consultant for KCRC during the development of the variable elements. These were :-

- Design, procurement and installation of electrically operated doorgear
- Design and installation of new, modern, cab front-ends
- Renewal of motor coach bolster beams
- Rewinding of main transformers, and supply of molecular sieves
- Design and installation of a new interior
- Design and installation of a Visual Passenger Information System.

The design process was of an iterative nature : GARMS would propose, KCRC would respond, with questions, alternatives and eventually agreement. Because time was short, this process often became fraught, and occasionally we were not given sufficient time to satisfy ourselves as to the efficacy of a particular idea, or we weren't happy with the proposals on material, for example. Some of these problems went on to exhibit themselves as design defects in the finished product, and will be subject to retrofit at some time in the near future. However, as GARMS were required contractually to put the first train into service 12 months after the Contract was signed, it is understandable perhaps why the process did not always work as it was meant to.

A part of the Contract worthy of further mention is that relating to IPR. The KCRC Board were concerned at having no option but to appoint GARMS to carry out the works, due to the technical risk issues described earlier. Single tender was further justified by employing the "open book" approach to giving visibility of costs. However, the Board did ask for something in return, and GARMS agreed to provide KCRC with a Royalty Free IPR Licence covering the existing EMUs. Additionally, the IPR of the consultancy design work carried out by GARMS on the variable works items also passed to KCRC.

The Contract determined that the variable works would be converted to fixed works after the conversion of the 3rd train (i.e. the 36th vehicle). The reasoning here was that GARMS would be well down the learning curve by this stage, and the labour estimates would be more realistic. In reality, although the price was fixed after the conversion of the 3rd train, the labour element can be revisited after the production of the 8th train, as it was evident GARMS was still at the high end of the learning-curve after producing train 3.

During 1996 and up to July 1997, a regular dialogue between the KCRC project team and GARMS saw the elemental prices slowly coming together. This was a nerve racking time, as costs would emerge which neither side had envisaged during the project estimates phase.

As we clearly had to work within budget, and the scope had increased due to the Main Transformer works, a degree of "value engineering" was required. A number of options were open to us :-

- Increase the amount of material sourced from S.E. Asia
- Reduce the scope
- Audit to ensure GARMS had not made errors in their price.

As around 80% of the cost of the GARMS works were material, we had jointly recognized at an early stage in the project that efforts must be made to secure quality material in the most cost effective way. Accordingly, efforts were made by GARMS to discover suitable suppliers within China, Hong Kong and the Asia Pacific region. This was done (not without difficulty), and savings of up to 35% over European supply obtained. Some 20% (by value) of the material for the refurbishment project has been obtained in the Far East, saving around £2,860,000. Of course there were risks involved in using these hitherto untried suppliers, and as GARMS was responsible for technical risk, it was agreed that they must have the final say on material sourcing. All savings from materials purchased from the Far East came to KCRC.

The process to fix the price was detailed and complex. Eventually all the bills of materials (BOM's) were agreed for content, and the auditing of material costs was concluded. Inevitably a degree of negotiation took place where the scope of the works was reduced slightly to reduce costs, and bring the fixed price within budget. This was concluded and the price signed off in October 1997. It was now up to GARMS to deliver to the revised Contractual Programme the remaining 312 vehicles before 20th Dec 1999.

7. THE STORY SO FAR

The contract with GARMS was signed at the end of October 1995. The Hong Kong based GARMS design and Project Management team took residence from January 1996, the KCRC Consultancy team arriving a little later in February 1996. Both teams worked very hard during this period, as there was a lot to do, and only a short time in which to do it.

Meanwhile in our Ho Tung Lau workshops, GARMS construction team was readying themselves for the arrival of the first 3 vehicles for conversion. These came into the workshop on 23 July 1996, and the physical works commenced.

We had agreed with GARMS that for the first 3 trainsets, all the materials would be sourced from the UK, with the Far East suppliers only coming on-stream afterwards. This approach, whilst necessary for the early trains, gave problems when the Far East supply was turned on, due to delivery and quality problems. However, with some of the UK suppliers also giving problems, coupled with GARMS's policy of carrying out inspections of UK manufactured items only in Hong Kong, we soon began to fall behind programme.

This was compounded by our belief that insufficient design resource was being allocated to the job, evidenced by the fact that the promised full design team never materialized in Hong Kong.

Finally in December 1996 the first 6 vehicles started test running, and were officially launched by our Chairman. The first 12-car refurbished train entered service on 17 Feb 1997, some 3 months late on programme, but still a remarkable effort (16 months from Contract signing) especially when one considers the extent of the refurbishment.

Passenger reaction was initially mixed, but positive comments have grown over time. The reliability has been very good, spoilt only by a few door and Passenger Information Display problems. The new Traction package has so far been trouble free. The only cloud on the horizon remains our inability to keep the floor clean!

Perhaps the best customer feedback is that most people think they are travelling on <u>new</u> trains!

8. CONCLUSION

This project delivered new trains for old, the first in 16 months, at less than 30% of the cost of a new train. KCRC's needs at the time could not justify new trains - refurbishment and re-engineering seemed the answer.

An assessment of the various attributes associated with a partnership approach are listed below :-

- A better understanding of the product by both parties. One manifestation of this has been the low number of variation orders generated.

- A symbiotic working relationship, based on an understanding of each other's requirements and restrictions. For example, GARMS understands KCRC's restrictions on availability and the need to link refurbishment with overhauls. Similarly KCR has been flexible in increasing the numbers of vehicles available to GARMS, together with storage facilities and expanded worksite needs.

- Working together exposes GARMS to the realities of the running railway : direct feedback of complaints to them following incidents, for example.

- Do conduct regular contract reviews, from the beginning of the project.

This project gave a supplier and a customer an opportunity to try a fresh approach to doing business. The paper, I hope, reflects the complexity of the project, and the challenge undertaken to conduct a major reconstruction of the fleet on one of the world's busiest railways, whilst maintaining normal service to the customer. The success of the partnership is based on the realization by both parties that they have to be flexible and mutually supportive, and the knowledge that it is necessary for both parties to succeed - the failure of one bringing down the other. Although the partnership has gone through some difficult times, with perhaps more in the future, one cannot deny that the advantages do outweigh the disadvantages, and the finished product is superb.

The author wishes to thank all those within KCRC and GARMS who gave assistance to enable this paper to be written.

References : "The Development of a Risk Based Approach to Ensuring Safety for an EMU Refurbishment Project", by M.W.Q. Yeoman and A.J. Smith - Asia Rail 1997, Hong Kong 20 March 1997.

C547/020/98

New maintenance packages – are they a good deal?

R WAHIDI MBA, BSc, MS, CEng, MIMechE, MCIT
Booz·Allen & Hamilton Limited, London, UK
P J EDWARDS BEng, AMIMechE
Booz·Allen & Hamilton Limited, Hong Kong

"NEW MAINTENANCE PACKAGES - ARE THEY A GOOD DEAL?"

RASHID WAHIDI MBA (UK), MS (USA), B.Sc (Engg) CEng, MIMechE, MCIT
Booz·Allen & Hamilton Ltd. (UK)
PETER EDWARDS B.Eng. (Hons), AMIMechE
Booz·Allen & Hamilton Ltd. (Hong Kong)

SYNOPSIS:

The train maintenance market is undergoing rapid changes with manufacturers offering many different kinds of maintenance packages with their new trains to meet the business needs of Operators and Lessors. In some cases the maintenance of existing fleets has been included. Many of these packages offer to take the risks of maintenance away from the train operator - but are they a good deal? Each Railway has its own special needs. The paper reviews these and sees how they impact upon the packages. What are the opportunities and risks for the operator, manufacturer, and lessor? What is prompting different railways around the world to be willing to pay for different levels of maintenance? What assessment criteria are used to evaluate maintenance package benefits?

1. INTRODUCTION:

In the UK and USA the rapid changes in the railways and transit authorities respectively have been the main driver for the development of maintenance packages. The privatisation of railways in the UK and local Transit Authority initiatives to reduce railway operating costs and improve performance in the USA, have caused operators and maintainers to look closely at railway processes not only on both sides of the Atlantic but also elsewhere in the world (1). Train maintenance is a primary operating cost and service delivery influence for all railways, and has therefore received significant attention over the past few years. Many railways are implementing fundamental changes. Within the UK, privatisation of the British Rail passenger businesses into twenty five Train Operating Companies was a major step change. These franchised train operators are answerable not only to their customers and the regulator, OPRAF, but also to their own shareholders for the bottom line performance.

Operators therefore desire to improve performance whilst significantly reducing their costs. Procurement of new vehicles potentially gives an operator the opportunity to achieve step changes in train reliability, performance, maintenance arrangements, customer satisfaction and profitability.

Initiatives are therefore now more focused on delivering improvements to:

- Revenues
- Operating costs
- Train performance
- Customer service

For many railways in Europe, the USA, Australia and Asia, Booz•Allen & Hamilton has analysed existing maintenance practices to identify (and in some cases implement) improvements consistent with the overall business objectives of the given railway, including procurement of new trains with or without maintenance options. This paper relates current trends in the areas of train maintenance, identifying business drivers, maintenance package options, and assessment criteria.

2. WHAT IS PROMPTING RAILWAYS TO GO FOR INNOVATIVE MAINTENANCE ARRANGEMENTS?:

Within many nationalised railways, and heavily subsidised metros, train maintenance engineers have worked principally on the basis of providing sufficient safe and reliable trains to meet service requirements. There was less emphasis on the processes employed to achieve this, the performance and the costs. With the current trends to improve railway profitability, the business case associated with train maintenance, investment and contracting strategy are becoming central drivers. An increasing number of maintenance organisations are tracking output based on key performance indicators related to cost and service delivery, and setting targets for performance improvements to deliver overall company objectives. Various techniques and continuous process improvements are now being adopted by maintenance organisations to ensure the achievement of the performance targets - one of which is the use of innovative maintenance arrangements.

Train maintenance in the UK can claim to be in the forefront of such developments, involving four key groups of players:

- Operators
- Manufacturers/Maintainers
- Lessors
- Regulators

The current influences on train maintenance are shown in Figure 1 below:

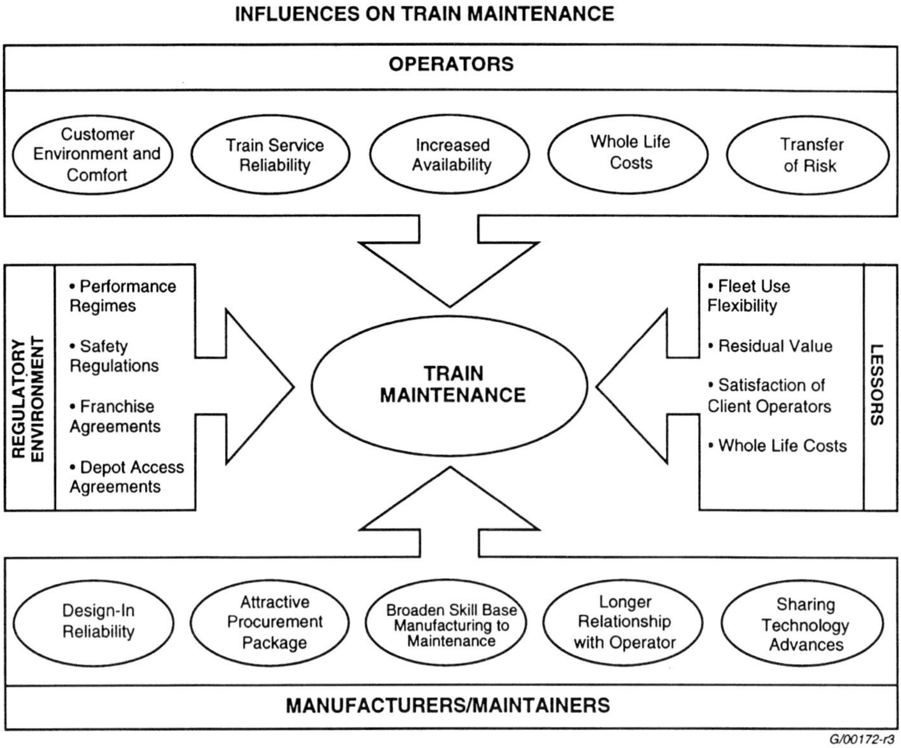

Figure 1

The key strategic business objectives of the private sector players now directly influence the principles and processes applied to train maintenance. These may include:

- Profitability
- Growth
- Market Share
- Return on Investments

Each player has defined its objectives as part of its business planning process, such that operations and maintenance organisations can develop principles and processes consistent with the company's vision and strategic objectives.

3. CHANGES IN THE MAINTENANCE MARKET:

The UK rolling stock maintenance market is undergoing rapid changes with the train manufacturers offering many different kinds of maintenance packages with their new trains in order to meet the objectives of operators and lessors. In some cases the maintenance of existing fleets has been included. Many of these packages offer to take the hassle of maintenance management away from the train operators allowing them to concentrate on service delivery and marketing.

The traditional method of the train operator purchasing the vehicle and then owning, maintaining and taking full responsibility for the vehicle, after the warranty period, for its entire working life is no longer applicable in the context of franchises of seven to fifteen years duration. Franchisees are also less willing to be responsible for the large risk associated with fleet ownership.

Virtually every vehicle built has experienced some unforeseen failures in its lifetime. After the initial warranty period the train owner is exposed to the financial and operating risks associated with equipment failure. The vehicle that requires no modification after delivery has yet to be built! Under the traditional method of procurement the owners and operators of the vehicles take on the risks associated with long term ownership, including design flaws which may not manifest themselves until the train is several years old. The modifications to correct these flaws can sometimes involve considerable engineering and operational effort and expense.

The trend is towards reducing risk and improving service delivery whilst reducing costs. This is not restricted to UK but is also at different stages of implementation elsewhere in the world. There is also a much greater awareness and change in thinking amongst rail operators worldwide as shown in figure 2 below.

CHANGES IN RAIL OPERATOR THINKING

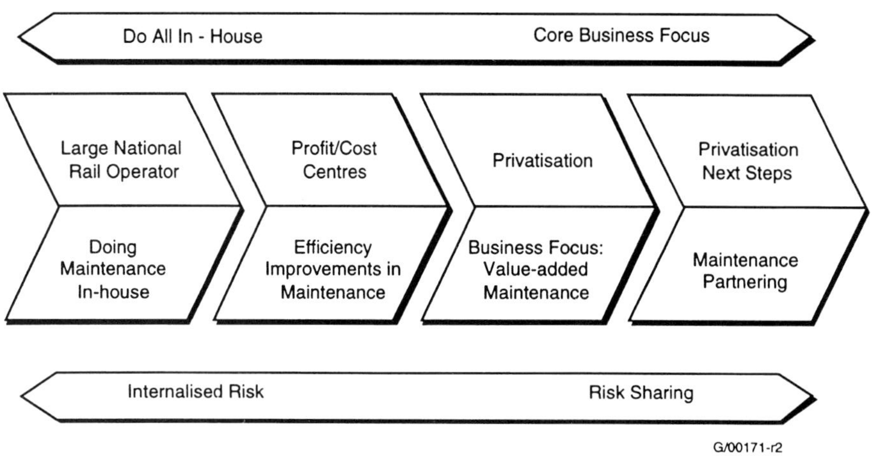

Figure 2

As can be seen maintenance has moved from the tradition of all work being done in-house by the operator. These changes briefly are:

Efficiency Improvements in Maintenance:
Fleet maintenance was basically managed within a budget based on past experience, which provided little or no incentive for radical ideas in maintenance practice. Expectation mainly to control cost and do efficiency improvements.

Business Focus: Value - Added Maintenance:
Each task of train maintenance is examined to see what value it adds to service delivery and the bottom line of the company as a whole. Incentive to innovate and rethink processes and performance while becoming increasingly cost effective.

Maintenance Partnering:
Maintaining trains of other companies or getting the trains maintained by a train manufacturer or maintainer. These are normally longer term relationships with a partnering approach including performance based relationships. Innovation based on partnering or novel relationships with risk sharing.

Train operators considering the option of contracting out their light maintenance responsibilities may find that it may not make sense for all, especially those with a large diverse fleet maintained at several different places. However for those with a relatively modern fleet that can interest the train manufacturer in particular, for those who are considering the purchase of additional new vehicles or a new fleet, the prospect of transferring risk and responsibility to an organisation better equipped to deal with these demands should be an appealing one (2).

4. OPERATOR NEEDS:

Different operators have different train maintenance needs, depending on their passenger markets, operating environment, fleet mix and age, depot locations and journey lengths. Most however, will require defined levels of train reliability, train availability, equipment reliability and customer environment to be maintained. Whatever the operators' needs, these will influence the targets which must be set and the maintenance arrangements which have to be made. For instance a long distance Inter city railway will define its punctuality targets in a different way to a metro or light rail system. Equally, they will define reliability and availability of the customer environment of their vehicles in different ways. It is important when specifying the requirements that the operator decides what his priorities are and makes sure that the contract conditions reflect these. If an operator wants his heating/air conditioning systems available and working correctly 100% of the time then this needs to be specified and poor performance suitably penalised.

If the train is required to operate through areas where vandalism occurs then this needs to be recognised. The sharing of risk for third party actions is critical in reducing the effects of vandalism. With the current maintenance contracts in the UK, the maintenance contractor bears little or none of the risk and cost associated with third party actions. This is despite the fact that a substantial amount of train maintenance budgets are spent repairing vandalism. If the contractor can share in the benefits associated with reducing this then it can be quite an incentive to work closely with the operator to make the train more vandal resistant. Otherwise there is little incentive for the manufacturer to build a train designed to minimise the effect of these third party actions.

Most contracts in the UK offer little protection from this however some maintenance contracts let in Australia requires contractors to bid based upon a projected level of vandalism. It is only when the amount of vandalism that has occurred is a certain percentage higher than this amount that the contractor can claim for a variation. If the amount of vandalism is reduced as a result of action by the contractor either through design or detection then the contractor is incentivised for the difference.

5. MAINTENANCE PACKAGES IN THE MARKET:

With the new maintenance contracts the operators and owners of the vehicles are transferring many of the risks associated with long term ownership to where they are probably best dealt with, the vehicle manufacturers. Consequently many variants of maintenance packages have appeared in the market, mainly related to new trains.

A design, build, maintain contract also allows the operator to transfer some of the risks associated with train equipment failures and non-availability, through performance penalties in the contract, as well as firmly leaving the responsibility for re-engineering any fleet wide problems with the train manufacturer.

Typical benefits of such contracts include:

- the train will be better designed for maintenance as the manufacturer will place more emphasis on maintainability

- maintenance costs will potentially fall as a result of the competitive tendering process

- components and systems are more likely to be selected on the basis of whole life costs rather than on a cheaper component basis

- maintenance manual quality is improved as a result of the manufacturer needing to use the manuals themselves

- the traditional problems of obsolescence, uncooperative original equipment suppliers and long lead times are removed from the rail operator and placed with the manufacturer who for the most part is much better positioned to deal with these problems due to ongoing new train build contracts

- suppliers gain more knowledge about the failures of their equipment and ultimately further improve the design of new trains and spares as contracts increase

- more effective sharing of information than would otherwise be the case

- contract administration is made considerably simpler because of reduced number of contracts or agreements between the operator and other parties. Typically separate contracts for train leasing, heavy maintenance, light maintenance and cleaning may be replaced by one contract. This reduce the interface risks and administrative burden.

Maintenance Packages on Offer

There are a number of different types of maintenance contracts now being used in various parts of the world. These include some of the types of maintenance arrangements as shown in Table 1 below:

Maintenance Package	Brief Description
Design, Build Operate Maintain	Consortiums build a complete rail system which it then operates and maintains it to a pre-agreed price and other performance based criteria .
Design , Build Maintain (1)	Operator orders train from manufacturer and subsequently negotiates a performance based maintenance contract with the manufacturer .
Design, Build Maintain (2)	As (1) above but with performance based maintenance contract specified in the original design, build and maintain invitation to tender.
Design, Build Heavy Maintenance	Operator procures train from manufacturer who at the time of procurement commits to undertake certain heavy maintenance. Standard required, timing of the heavy maintenance in the trains life, and costs are agreed at the start.
Specification based Contract Maintenance	Operator owns existing vehicles and contracts maintenance to a maintenance contractor. Operator specifies in detail the standards/schedule to which vehicles must be maintained.
Performance based Contract Light Maintenance	Contractor performs light maintenance on train on behalf of the operator who may or may not be the owner. Operator specifies performance based specification. Separate contracts may be in place for heavy maintenance.
Life Cycle Spares supply	Train owner procures vehicle from supplier with an undertaking to pay the supplier a fixed amount each year for spares supply. Amount is determined by supplier at bid stage of contract for vehicle supply.
Maintenance Partnering	Creation of a joint venture company or other partnering arrangement between operator and manufacturer for train maintenance.

Table 1

Design Build Operate Maintain

The Design Build Operate Maintain or DBOM as it is known is becoming increasingly popular in areas where a government or local authority wishes to finance improvements to the transport infrastructure but does not want to manage the day to day affairs of the new railway. Instead it provides some of the capital to build the new railway but lets a concession for the operation of the railway. This is a common type of contract popular for new Light Rail projects in many parts of the world.

The government lets a contract for design, build, operation and maintenance of the railway but is still responsible itself for determining service levels, control of revenue, and long term strategic decisions affecting the railway.

Design, Build and Maintain - Performance Based

This is now becoming increasingly popular as a method for procuring new trains in the UK and abroad. The contract for these typically specifies availability and reliability requirements for the rolling stock and provides standards to which the rolling stock must be built . The responsibility for meeting these requirements then becomes wholly the responsibility of the contractor. Some contracts require the contractor to determine fleet size , some give a minimum fleet size that they would expect to see the contractor provide. All contracts place penalties on the contractor for failure to meet the specified requirements.

The development of the performance specification for this type of contract is critical. Generally the higher the levels of performance required, the more the contractor will want to mitigate the risk of not meeting the performance requirements . This can manifest itself either as an increased cost or as protracted investigations into the root cause of every failure as the contractor attempts to place some of the blame for the failure back on to the rail operator. The cost effective contracts are ones which set targets which the contractor recognises as achievable, but which are still an improvement over the service currently being offered.

To achieve optimum value for money and risk transfer, the maintenance arrangements must form an integral part of the competitive tendering process for the vehicles.

Design Build Heavy Maintenance

This is the DBM contract at the very simplest level. At this level there is limited compensation for the operator if the vehicle is not returned to service on time and there is no compensation or accepted liability for in service failures. The rail operator determines the availability of the rolling stock for maintenance and also the size of the fleet required for service. This type of contract may have some minor enhancements to provide some level of incentive to the contract maintainer. This option has the benefit of the simplest contract administration arrangement as well as providing heavy maintenance efficiency improvements.

Specification based Contract Maintenance

Operator owns existing vehicles and contracts maintenance to a maintenance contractor. Operator specifies in detail the standards/schedule to which vehicles must be maintained and audits/inspects the work on a regular basis. This type of contract has been in use for a while mainly on existing fleets.

Performance based Contract Light Maintenance

Contractor performs light maintenance on train on behalf of the operator who may or may not be the owner. Operator specifies performance based specification for the maintainer. Separate contracts may be put in place for heavy maintenance if desired. May be on existing fleet or new build.

Life Cycle Spares Provision

This is a relatively new concept. With this type of contract the rail operator maintains the vehicles but some of the risks associated with lifetime reliability of equipment and issues such as obsolescence remain with the Supplier.

Effectively the supplier is required to provide a life cycle tender for each of the selected components and systems on the vehicle. This fixed price will include all replacement spares, and could be paid on an annual basis, independent of spares usage. The incentive therefore remains with the supplier to continue to implement product improvements.

This approach does not directly encourage the supplier to address failures of inexpensive components which may be reducing service reliability, so the supply contract will still need to specify service reliability separately to cover this. However this method does provide many of the benefits of design, build, maintain contract. Not least of which is the reduction in spares usage.

Maintenance Partnering

Creation of a joint venture company or other partnering arrangement between operator and manufacturer for train maintenance. New development in the industry which is currently being discussed and may be implemented in the UK.

6. ASSESSMENT CRITERIA:

Assessing the benefits of a train maintenance package is not easy as it involves many different issues. The key criteria are discussed here:

Punctuality

It is essential that the operator has a good understanding of what he means by punctuality and how it is influenced to different degrees by, not only the failure of the train but also the responsiveness of those driving or maintaining the vehicle. This will typically, but not always, involve the use of an 'impact minute' type technique to measure punctuality. This type of measurement will generally work well provided that it is being applied in a fair. pragmatic and consistent manner. Too strict an adoption of this technique can potentially have negative effects as the operator and contractor may fail to agree on the cause of failure and the debate about measurement of performance may hinder overall improvement.

Availability

Availability is possibly one of the most used measures in delivering performance based contracts and is also possibly one of the most misused. The definition of availability varies from one contract to another and for commuter based operators, the concentration is on ensuring that the peak service is delivered. Whilst this is a laudable goal, when it becomes the focus of an organisation it can have questionable value. For an inter city operator the availability objectives may include increased utilisation of the available trains as well. Indeed, one train operator is achieving 100% availability in operation at certain times of the day. Too heavy an emphasis could potentially reduce the overall reliability of the fleet by encouraging the contractor to run trains with intermittent or incipient subsystems faults.

Reliability

Reliability of the train as well the main sub systems and equipment is an important criterion for the overall train performance and service delivery. Reliability is measured differently by different operators as mean distance between failures (MDBF) or mean distance between significant failure (MDBSF) or mean equipment service hours before failure. These are normally technical measures and may be prioritised after the punctuality and availability.

Customer Environment
This is one of the more difficult areas to assess, and can often lead to disputes between the client and contractor. There are three commonly used techniques too assess performance. These are (3):

- inspection by inspectors to agreed visual standard
- customer satisfaction surveys
- 'mystery shopper' surveys

Setting targets in this area is a relatively simple task for trains that are already running on a railway. Any improvement or worsening of performance can be easily detected. With a new fleet of vehicles, determining and setting a standard to be achieved can be difficult. It is important that the method employed uses a consistent measuring technique and that any change in performance of the contractor is readily detectable, but also takes into account any changes in customer expectations after the stock has been delivered.

Spares Holding
The spares holding for a fleet of vehicle will typically reduce as a result of the fleet being maintained by the manufacturer. This is for a number of reasons :

- if the component is used in a number of fleets the manufacturer may choose to hold less used spares centrally. This reduces the number of parts on site
- the desire to purchase more spares initially is reduced as the maintainer now has greater leverage over the OEM than the operator did previously

Obsolescence
Protection from obsolescence is one of the main reasons why operators look to have their vehicles maintained by the original manufacturer. To protect themselves, most train operators traditionally will over order the initial spares, increasing the initial investment costs. This then increases spares holding. When the initial order of spares runs out the operator will frequently experience significant problems. The original supplier may not want to know about manufacturing a component that is more than five years or so old unless it is still part of their catalogue. At best the operator may be quoted a very long lead time and a very high price.

As a result of many years of experiencing obsolescence, long lead times and high spares prices from OEMs operators build up teams of people skilled in the black art of 'reverse engineering'. Contracts requiring the manufacturer to make long term spares provision, or do a partnering arrangement in maintenance, pass on this responsibility to the manufacturer and the operator now no longer needs 'reverse engineers' and is potentially not exposed to the risk of obsolescence.

Cash Flow
The cash flow in general and of the train maintenance function in particular is a key aspect of the present day business environment in the railways. The maintenance cash flows in the past have been 'peaky' due to the unpredictable nature of the maintenance activities. With the maintenance packages, the cash flows are now predictable and in typical arrangements would be reviewed on a period by period basis. The planning for the peaks or unpredictability of the future maintenance activities is a risk transferred to the manufacturer or maintainer.

Capital Costs
These mainly relate to the capital investment necessary in the maintenance activities or depots:

- investment of capital in essential spares by the operator so as to reduce the risk of vehicles remaining unavailable due to non-availability of spares
- investment in diagnostic equipment
- additions to the maintenance facilities at depots

These capital costs are transferred to the contractor either partially or fully in the maintenance packages.

Ease of Measurement:
The more complex the measure, the more difficult it will be to do the measurement. It is therefore useful to keep the maintenance package output measures simple.

7. CONCLUSION:

It is too soon to say how successful the manufacturers/maintainers will be in delivering their promises of overall benefits to the operators. So the question 'maintenance packages are they a good deal' cannot yet be completely answered.

Maintenance packages do have the attraction of taking the fleet and depot management issues away from the operator, who pays for a service and gets the desired train maintenance as per his train operation plan. The packages certainly have the 'hands-off' appeal. However, traditionally the skills of train maintenance have been with train operators staff. The cost benefit assessment really is the factor of 'are we getting value for money?'. In other words what is it costing the train operator to pass on this part of his work to a contractor both in terms of cost and performance enhancement, and risk transfer.

As the market is developing and opening up, this area will become more and more competitive. In addition to commercial benefits, there will be other longer term, and less tangible, benefits to the railway industry as a whole.

Some concluding thoughts:

Operator's perspective:
- it is the most successful way train operators have found in securing a long term commitment for the product from the manufacturer
- maintenance packages are attractive to franchised train operators particularly those ordering new trains
- operators will continue to maintain most of the existing fleets in the near term but as the market develops, more attractive maintenance arrangements for existing fleets will follow
- economies of scale impact the decision
- operators have to pay for the risk transfer but can never transfer all risks (e.g. loss of franchise)
- a partnering and win-win attitude and arrangement needs to be in place otherwise success may not be achieved

Manufacturer/Maintainer's perspective:

- manufacturers expect to build a profitable business from the maintenance service but also intend the operations to provide a shop window for their services and skills
- largest manufacturing groups have made the strategic decision to provide maintenance services and are offering maintenance package deals - mainly with new train orders
- maintenance package closes the loop and transfers the risk to a party which is now prepared to take a long term view of its products
- a partnering and win-win attitude and arrangement needs to be in place otherwise success may not be achieved

References

(1) C Hayward-Williams and R Wahidi , 'USA Maintenance practices - do they know better?', paper C530/028/97 IMechE Conference, 'Train Maintenance Tomorrow and Beyond', York 1997

(2) G Daniel and AI Hamilton , 'Subcontract Maintenance in the Privatised Railway', paper C530/031/97 IMechE Conference, 'Train Maintenance Tomorrow and Beyond', York 1997

(3) RG Consterdine , 'The development of Cleaning Contracts', paper C530/032/97 IMechE Conference, 'Train Maintenance Tomorrow and Beyond', York 1997

(4) I Flynn, 'Train Maintenance by the Manufacturer - from business concept to practical reality' paper C511/19/105 IMechE Seminar, 'Design, reliability and maintenance for Railways', Railtech 96 Birmingham 1996

C547/019/98

LCC – what is the real impact upon design?

M THOMAS MSc, FIMechE, MIEE, **L de VLIEGHERE,** and **L de COEN**
Bombardier Transportation, Brugge, Belgium

LCC, WHAT IS THE REAL IMPACT UPON DESIGN?

1. FIRST SOME HISTORY: GROWTH IN LCC REQUIREMENTS.

As with all complex and high level functional requirements, the origin comes from the military and the aeronautical industry. The introduction of reliability and LCC into the railway industry was by adapting methodologies developed for these two industries

One of the first large scale applications in Railway industry seems to have been the X2000 train supplied to SJ, the Swedish national railways in 1990. This application has been very successful. Some publications explaining the approach followed for the X2000 train are available ([1]). They focus however on the contractual aspects and on the verification of the LCC, less on the work performed during the design.

Since the X2000 experience, the rolling stock specifications have gradually started to contain reliability and LCC requirements. We are now at the point that all specifications contain LCC requirements, all be it at different levels of significance.

The questions are then:

1. Has it made our trams and trains better?
2. Are they more reliable, easier maintainable?
3. Is the maintenance cost lower?

And if the answer to these questions is yes, how has it been done? We want to give some answers to these questions, especially from the engineering point of view.

2. DIFFERENT RAILWAY ADMINISTRATIONS: DIFFERENT APPROACHES WITH RESPECT TO LCC REQUIREMENTS

As many Heavy Rail and Light Rail operators as you have in Europe, there are as many approaches you have with respect to LCC definitions and requirements. Trying to explain all these different approaches in a paper such as this becomes impossible. An overview of some of the basic approaches is shown below.

In most cases large national heavy rail operators are rather conservative with respect to LCC requirements. They have little or no requirements. In some cases they ask in the tendering phase for a LCC calculation, this calculation is then used to compare different competitors, but seldom is LCC warranty asked for. Or the operators ask for a maintenance scheme. The operators have several reasons for this approach.

- They have their own engineering department. They are able themselves to judge in detail the technical solutions and the associated reliability and maintenance cost.
- In many cases they prescribe in detail the technical solution that has to be offered. In this case all competitors have to offer the same equipment, and of course this results in the same LCC
- In some cases they are also afraid of the notion LCC. It is no secret that most national railway operators still have too many staff and maintenance is still done in the traditional ways in their own workshops and has not been optimised with the result that in some cases detail costs are not known, only global values (e.g. at depot level) are known. The optimisation of maintenance procedures, which is the logical consequence of a LCC approach, could result in some people becoming redundant.

Another approach is that of the LCC guarantee. In this case the rolling stock supplier has to guarantee the LCC during a certain period of time. When the LCC is exceeded the supplier has to pay a penalty to the operator.
This approach has the advantage that it will oblige the rolling stock supplier to think about reliability, maintenance cost, maintainability, ... during the design phase. This will inevitably lead to a better vehicle. On the other hand there are also some disadvantages.

- During commissioning and operation of the vehicles a lot of effort is spent in proving that the LCC is met / not met. The LCC demonstration can become very difficult and time consuming; this effort could better be spent in solving the technical problems. There is no win-win relation. In this case a pragmatic approach from both sides is very welcome.

- For the supplier there is the disadvantage that he can only loose money. In the case that the LCC is not met, the supplier then has to do corrective actions and/or to pay penalties. However, some contracts provide for bonuses in the event that LCC is better than contractually agreed.

The above approach is followed in general by public operators, who have their own personnel which they find difficult to make redundant, but who have on the other hand "paid by performance" contracts (e.g. paid per passenger-kilometre) with their mandatory organisation. They take this approach to be able to rationalise their operation and especially their maintenance cost. Some Dutch light rail operators have adopted this approach [2].

The third approach is that of the maintenance contract. Here the rolling stock supplier also supplies the maintenance at an agreed cost. This approach has the same advantage as above approach but not its disadvantages.

- In case of technical problems (e.g. systematic failures, maintainability not good in certain areas...) efforts will be made by the supplier (who is responsible both for doing the maintenance and for the cost of it) in solving the problem, not in arguing about the responsibility.

- For the supplier there is a trade off. In case something goes wrong in one area, he can and he is motivated to optimise in other areas. Due to the trade off he also has the means to do so quickly and in an efficient manner.

This approach is typical of the new privatised operators in the UK

Other approaches with respect to LCC requirements are possible. In most cases they are a mixture or a variant of the above 3 basic approaches.

3. SOURCE INFORMATION AVAILABLE

When you have to tender or to design a vehicle for LCC, you first have to know your starting point.

Question 1: what is the actual LCC of similar vehicles and where is it available?
Question 2: what are the weak points of the existing designs, how can I design them out and how much will it decrease the LCC.

A third question adds to the two above.

Question 3: what is the accuracy and sensitivity of all these data and estimates. In other words what is the risk associated to these data.

Question 1: what is the actual LCC of similar vehicles.

Here of course, we as rolling stock suppliers first have to blame ourselves for the lack of information. For decades we have supplied vehicles and have showed little interest in their further use and maintenance, afraid as we were to have to intervene to solve failures. This being said, it is very difficult to get LCC data from the operators.

- Some operators just have no data. Other operators don't want to supply it, afraid as they are to show their rather low productivity.

- Other operators can only supply very general data, e.g. the total yearly maintenance budget, or the total budget of a certain workshop. This is very difficult data to deal with; even if you know the number of vehicles maintained by the workshop and the number of employees, the calculation of the net maintenance cost of the vehicles is impossible, because of many variables for example vandalism and accidents, the productivity rate, the amount of overheads and logistics, the amount of work outsourced, ...

- Some operators can supply data per maintenance action and per main sub-system (propulsion, brakes, doors...). However care has to be taken: for example are the indicated man-hours net working time, or does it include waiting time, productivity... To give an example: a certain repair could take 2 hours on one day, and the same repair could take 8 hours another day, just because the first day there was a lot of work and the second day the employees had nothing else to do.

- The most effective manner to get detailed data per maintenance action and per main system is to do this yourself in close co-operation with the operator and to do follow-up in the workshop of the operator during a limited period of time.

Question 2: what are the weak points of the existing designs, how can I design them out and how much will it decrease the LCC.

Once the detailed information per maintenance action and per sub-system is available the weak points can be detected such as items costing a lot of money to maintain. So for instance you will find on existing designs that:

- DC traction needs a lot of maintenance i.e. change the motor brushes.
- some systems have a lot of brake shoe wear, due to a bad performing electro-dynamic brake
- piston type compressors are rather reliable but need frequent overhauls compared to screw compressor
- etc...

Once a number of such items are identified, they should be compared with new technology and the improvements in LCC quantified. Examples are:

- AC traction has no motor brushes and no current collectors, ...
- there are good electro-dynamic brakes on the market leading to long life of the brake shoes
- rotary vane or screw compressors don't need frequent overhauls

All these technical improvements correspond to a decreased maintenance cost, which can be calculated. But has the new technology any drawbacks, which will lead to added maintenance cost in other areas?

Question 3: what is the accuracy of all these data and estimates. In other words what is the risk associated to these data.

Risks are many:

- very little operators do a detailed follow-up of their maintenance cost, certainly not in Europe. The obtained data on sub-system level or on maintenance action level always requires interpretation.
- a vehicle or technology functioning well in environment X does not mean that it will be OK in environment Y (example: Channel Shuttle trains which suffer seriously from sand and salt which in projected on their underframe, something which was never experienced on other trains)
- new technology: usually developed to take away drawbacks of old technology, but sometimes creating new problems, leading to increased maintenance cost

4. CASE STUDY: THE ROTTERDAM SNELTRAM EXPERIENCE

4.1. The contract

In the spring of 1996 Bombardier Transportation signed a contract with RET Rotterdam to supply 36 Sneltram vehicles. The contract included a guarantee on availability and on LCC.

If the contractual availability is not reached on the fleet 3 years after delivery of the first vehicle, Bombardier Transportation has to supply extra vehicles for free. The LCC guarantee includes the cost of energy, cleaning of the vehicles and maintenance. A 7 year period is guaranteed. Where LCC are higher than contractually agreed at the end of the period, Bombardier Transportation pays the difference([2]).

4.2. The agreed values

The contractually agreed values are based on the actual performance of another similar Bombardier Transportation vehicle, the Docklands Light Railway, which is performing very well. However a number of potential improvement were taken into account in order to be able to offer to the customer an attractive vehicle.

4.3. Basic principles of the vehicle

Use of new technology has been limited.

> Only where needed new technology has been used: e.g. AC propulsion equipment with IGBT inverter modules. However also in this case it has been proven as much as possible: so will the propulsion equipment be proven, because exactly the same units are used on the Amsterdam S3/M4 vehicles.

For most equipment: no revolution but evolution.

> This means that as much as possible existing proven equipment has been used. This equipment has however been scrutinised rigorously with respect to its operating

performance. Any weak points have been designed out. This must guarantee us better equipment, without the risk of total new failures, which you inevitably have when changing of technology.

Some examples:

- The doors will be electrically operated sliding-plug doors. The same door system from the same supplier has been recently used on several previous orders by Bombardier Transportation with good results. However the supplier and Bombardier Transportation have worked together with each new order to further improve the door, without making any drastic change to the mechanism.
- The compressor will be a vane machine. The reputation of rotary machines on rolling stock is not always very good: oil consumption, oil-emulsification,... Therefore, all known problems of such machines have been identified (with current users) and been designed out by the supplier.

4.4. Design for low LCC and high reliability

Learn from the past.

During the tender and during the first stages of the design of the vehicle Bombardier Transportation has collected a lot of information on the operational performance of similar systems and sub-systems. This information was both quantitative and qualitative. We collected:

- levels of reliability achieved
- weak points with respect to reliability, technical analysis of those weak points
- problems with maintainability: difficult to repair, time-consuming to diagnose, difficult to access, ...

Information was collected from our own customers operating Bombardier Transportation vehicles (like DLR, Amsterdam Sneltram, Köln K4000, ...), but also from companies operating non-Bombardier Transportation vehicles. To the latter we didn't ask global information, but some specific questions on the functioning of certain sub-systems (like the brakes, the compressors, the heating-ventilation, ...). Indeed, most operators find it difficult to answer global quantitative questions (on R-levels, on LCC), but are in most cases very helpful if you ask qualitative detail questions. Their comments help identify actual weak points of designs, which can then be feed back to your sub-systems suppliers to make them improve their system.

A similar approach was followed for the fitting out, were we collected and documented a lot of information from our after-sales service, in order to feed it back in our design office during the design phase.

Independent verification of the design

All historic field information was collected by an independent group within Bombardier Transportation engineering, the RMS group. It was fed back to the engineering department and to the sub-system suppliers. Then on regular occasions, the design was reviewed for LCC together by the designer and the RMS group. Questions were asked on maintainability, accessibility, robustness, ... All historic experience was cross checked against the design. This lead to a very large number of mostly small changes to the design. All decisions for improvement were taken by the designers themselves, because they realised they would improve their system. A good and friendly co-operation between designers and RMS group is needed to obtain a good result: both together make a good vehicle.

A similar procedure was followed with the sub-system suppliers where their design was scrutinised with historic information and improvements were suggested. Some suggestions were followed, others not; after all the supplier remains responsible for their design.

Involvement of the suppliers.

All major system suppliers have taken over similar contractual obligations as Bombardier Transportation does for the main contract, but it does not stop with the contractual obligations. As described above the sub-system suppliers and Bombardier Transportation have worked technically together to collect historic field data and to design weak points out. Indeed, it does not serve any case to agree stringent LCC and reliability targets, and then sit down and do nothing and hope your equipment will reach the targets.

Independent assistance

The NS (Dutch Railways) are one of the European railway operators who have made over the last decade the most progress in optimising their maintenance. They have a lot of experience in maintenance research and have been able to reduce significantly the maintenance cost of their equipment.
Their experience was used on some dedicated areas like the pneumatic brakes and the bogies.

Maintenance manual and training.

Of prime importance are good and accurate manuals. It is a good thing to start making the manual whilst designing the vehicle; it provides another way to check the maintainability of your design.
Also the diagnostic part of the manual is important. It must be possible to start from the symptoms (which can be both announced by the vehicle system or just visually observed) to bring you in a logical manner to the defective module or component. Accuracy has to be high, to avoid replacing the wrong component.

5. THE ROLE OF DIAGNOSTICS IN REDUCING LCC

5.1. The targets of train and fleet management system.

For modern complex trains, safety, availability and life-cycle costs are directly related to the efficiency in fault diagnosis throughout the life-cycle.

Therefore during the design phase of a system/train different service and maintenance aspects have to be taken into account:

- Intrinsic safety: systems have to be designed to be safe. This implies that systems failures that can endanger the intrinsic safety have to be supervised and visualisation/reporting has to be foreseen for service and maintenance personnel.
- Comfort: systems which are providing a certain level of comfort to the user have to be designed to keep this level as optimal as possible and diagnostic information has to be provided when the level decreases.
- Availability of functions/systems/trains : to be able to guarantee a high level of availability redundancy and degraded modes are implemented where necessary, as well as possibility to take over certain functions by other means, e.g. by remote/local/external control. Other aspects that influence availability are the correct evaluation of service impact of system failures, taking the correct decisions and doing the right actions at the right moment and minimising the number of sleeping failures([3]).
- Maintainability: appropriate diagnostics signals have to ensure that the failures can be localised in an easy way and they have to be related to Line Replaceable Units, of which the accessibility has to be optimal.

5.2. Diagnosis Implementation

System/train diagnosis has to be implemented and integrated at different levels and serving different purposes ([4]).

A. Diagnosis on-board of the trains.

Embedded in the system.

- Each system with a certain level of intelligence (process capacity) performs auto-diagnosis. This can include memory checks as well as verification of the correct functioning of the input/output channels.
- A second level of in-system diagnosis is the process diagnosis, which validates the behaviour of sensors, actuators, processes, timing, ...
- Another possible source of information is available when start-up/shut-down tests can be executed.

In the Train Management System

- An important role of the TMS is the "alarm & fault" management. This includes the visualisation of detected problems for the driver, train captain or maintenance personnel on board of the train, but also the storage of the diagnostic information (localisation, problem identification, time of occurrence, ...) and the transmission of this data by a radio-link if available (otherwise it has to be done by paper of orally).
- Another role of the TMS is the service advice that an on-board integrated expert system can give to assist the driver to find the best action to be taken in case of a system failure or the automatic decision making for certain intersystem interactions.

B) Diagnosis in the control/maintenance centre.

All diagnostic information memorised in the TMS on each train can be transmitted to a central system in the control centre.
There an expert system can be used to assist the drivers in decision making for complex or new cases or for decision on fleet level.
The available train-information/diagnosis allows the fleet-manager to know the location, the configuration and the health of each train. With this information an optimal scheduling of the maintenance can be made, so that the fleet health can be managed to guarantee maximum safety and availability.
For maintenance the transmitted train-diagnosis can be used to define priorities and to organise the different necessary resources for the maintenance to be done for a train still in service. During the maintenance phase, an expert system can assist the operator in fault finding (for unique, complex or intersystem problems) and it can give the necessary repair advice (spare part, repair procedure, ...).

C) Other advantages

The system knowledge needed for operating and maintaining the trains is in the systems on-board and in the control and maintenance centre and all information is always available for all the different users.
All problem detection, reporting and solving is done in a uniform way. This can be used to give feedback to the system designers for future improvements.
The use of interactive systems on-board and in the control and maintenance centre allows to train people better and with real and complex situations which can be simulated in an easy way.

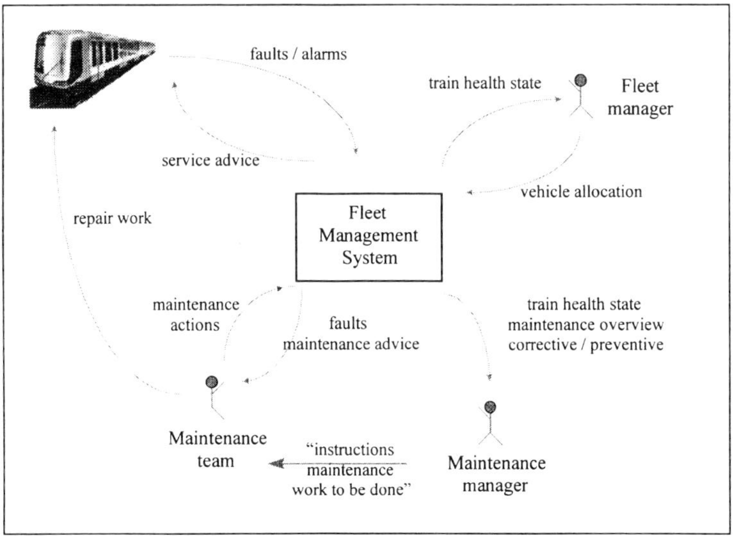

Figure 1

6. GENERAL CONCLUSIONS

- Designing for Life Cycle Cost requires a common sense approach during design: one has to collect field information from actually running vehicles and to couple it back to design.

- Good designed on-board systems interacting with an on-board Train Management System and an ground based Fleet Management System with expert system functions can improve in an important way the safety, comfort, availability, maintainability and thus the Life Cycle Cost of modern trains.

- The Rotterdam vehicles are not yet in service, so the acid test yet has to come: is the vehicle really better than its predecessors? Only reality can prove it. We can only say: come and see next year.

7. REFERENCES

(1) Akselsson M, Burström B. (1994). Life-cycle cost procurement of Swedish State Railways' high-speed train X2000. Proceedings Institution Mechanical Engineers, Volume 208, Part F: Journal of Railand Rapid Transit, pp. 51-59

(2) RET (Rotterdam Electrische Tram), Bestek nr. 113, vervanging van metrovoertuigen RET

(3) Coen L. De, 1991, Amélioration de la disponibilité au moyen d'un système d'aide au diagnostique embarqué, CEP SYSTEMES – Maitrise des risques dans le transport ferroviaires, 3.

(4) Netten B.D., and R.A. Vingerhoeds (1995). Large-Scale Fault Diagnosis for On-Board Train Systems, in: Case-Based Reasoning, Research and Development, (Eds.) Veloso M., Aamodt A., Lecture Notes in Artificial Intelligence 1010, Springer Ver

This paper was presented at an IMechE seminar Fault-Free Trains – A Reality? organized by the Railway Division of the Institution of Mechanical Engineers, held on the 1 April 1998.

C547/002/98

A wider UK market for simpler, titling trains?

C P BOOCOCK CEng, FIMechE, MIMgt, MCIT
Consultant, Mott MacDonald Limited, Derby, UK

SYNOPSIS

U.K. train operators tend to think of procuring tilting trains for high speed routes only. This paper offers suggestions as to how tilt technology can be applied to trains for other routes for which the advantages of tilting trains have not previously figured in procurement planning.

The paper goes further by proposing a more radical approach leading to a universal train design using articulation and component simplification. This approach would reduce the number of key components, would improve train reliability, allow better adhesion and better regenerative braking, and would reduce the overall price for train procurement and operation.

1. INTRODUCTION

Train operators, lessors and suppliers are seeking passenger train designs of universal application that can offer improvements in performance for less overall cost. This paper discusses development of two concepts:

- active body tilting; and
- articulation with component simplification,

and brings them together in a vision of a truly universal train concept.

2. TILTING TRAINS

2.1 What is a tilting train?

For the purposes of this paper I define a tilting train as one in which the passenger carrying vehicles are suspended in such a way that the carriage bodies are capable of being positively rotated through a few degrees from the vertical about a point near the body centre of gravity, so that passengers feel considerably less of the lateral force generated as the train rounds a curve.

Trains where each carriage body tilts as a pendulum due to the force exerted by the body itself as it rounds the curve are known as passive tilting trains. Such a train is the Spanish Talgo concept.

A train where the tilting is actuated by a powered mechanism in response to inputs from, for example, accelerometers and gyroscopes has active tilt. Most European tilting trains are of the active type.

2.2 Characteristics of tilting trains

Modern designs of active tilting trains are tending to converge towards trains with the following characteristics:

- active tilting mechanism powered by hydraulic or electric rams
- tilt angle decided from parameters measured by accelerometers and gyroscopes on train
- maximum angle of body tilt: 6° to 8°
- equivalent additional tilt from bogie secondary suspension: 2°
- maximum rate of change of tilt angle: up to 10° per second
- increase in train speed round curves (when compared with non-tilting trains): 25% to 30%
- effective compensation for lateral force on curves experienced by passengers: 70%.

It has been established that curving speeds of as high as 30% above those of conventional trains are possible with tilting trains. Generally, the advantage of tilt is felt on any curve that restricts a conventional train's speed, irrespective of what that speed is. For example, a curve that restricts conventional trains to, say, 50km/h may be traversed by a tilting train at 65km/h; a conventional 80km/h curve may be traversed at up to 105km/h, and so on. Thus, the benefit of tilt increases in relation to the length of any route that consists of curved track and is rarely dependent on the actual curve radii, except on curves of very large radii where a tilting train's own maximum permitted speed may limit any benefit from tilt.

Routes such as the east coast main line that only have occasional stretches of speed restricted curvature benefit less from the potential of tilting trains than do routes that have considerable lengths of sinuous track (e.g. Carlisle-Glasgow). It follows that the journey time benefit is greater on routes that have frequent curves that cause conventional trains to brake and accelerate many times during their journeys.

A passenger in a tilting train traversing a sinuous railway route should be relatively unaware of the curvature through which the train is passing.

A tilting train can pass through curves faster than a conventional train, within the safety limits of the track/train interface characteristics. In most existing applications, the gap between maximum comfortable speed (for passengers) and maximum safe speed (avoiding overturning) is more than sufficient to ensure train operational safety. Knowing that the safety margin is nonetheless lower than with conventional trains, most operators would wish there to be some automatic indication to the driver of permitted train speed at all locations.

An informed contribution to the discussion at the I.Mech.E. Railway Division conference *Train Maintenance Tomorrow and Beyond* at York in 1997 suggested that tilting trains should cost about 10% more than the equivalent conventional train of otherwise identical specification.

2.3 Purpose of tilting trains

All existing applications of tilting trains have one clear purpose, namely to reduce journey times by maintaining higher continuous speeds over sections of route. Applications in Europe have been in two distinct market segments:

- high speed long-distance trains, normally to avoid the construction costs of new railways designed for high running speeds; also to permit faster transits through hilly regions
- local services in hilly areas.

Fig. 1: Cisalpino Pendolino high speed tilting train passing through Alps

Fig. 2: German type VT610 tilting diesel unit for services in hilly areas

Tilting trains, by virtue of the shorter journey times they make possible, offer train operators the following potential benefits on routes with significant curved sections:

- increased patronage and revenue (due to more competitive journey times)
- lower energy costs (due to more constant running speeds)
- better vehicle ride (due to active elements in suspensions and to lower perceived cant deficiency)
- earlier turnrounds of trains giving possibility of smaller fleet sizes

2.4 U.K. opportunities

At present, two train operating companies in the U.K. have expressed interest in acquiring high speed tilting trains to operate on lengthy main lines on what we used to know as InterCity services. Many other train procurement projects have begun, none of which is expected to result in tilting trains being specified. This paper expresses surprise that this should be so.

Opportunities exist in numerous places on the U.K. railways for the introduction of titling trains, mostly in the form of otherwise conventional diesel or electric multiple units or unit trains. Some of these routes are listed here to open up the debate; there are many more opportunities than these:

- St. Pancras - Nottingham/Sheffield - Leeds
- (Waterloo) - Basingstoke - Salisbury - Exeter
- (Waterloo) - Southampton - Bournemouth - Weymouth

- (Waterloo) - Woking - Portsmouth Harbour
- Victoria - Arundel - Portsmouth Harbour
- Charing Cross - Tonbridge - Hastings/Dover
- (Paddington) - Reading - Taunton - Exeter - Penzance
- (Paddington) - Reading - Worcester - Hereford
- York - Leeds/Sheffield - Birmingham - Bristol - Exeter - Torbay/Plymouth
- Bristol - Weymouth/Portsmouth
- Liverpool Street - Norwich
- Glasgow/Edinburgh - Inverness/Aberdeen
- Inverness - Kyle/Wick/Thurso
- Carlisle - Glasgow/Edinburgh
- Carlisle - Kilmarnock - Stranraer

I take just one of these routes, Waterloo to Portsmouth Harbour, for brief evaluation as an example.

2.5 Smaller fleet for same work

The current base timetable for the two fast and semi-fast trains an hour between London Waterloo and Portsmouth Harbour requires a minimum of 7 trains in service (this study excludes peak services, which could also benefit). These trains are each normally composed of two four-car electric multiple units though they are extended to twelve cars at peak times and seasons. The route is relatively straight between Clapham cutting and Woking, but is very curvaceous between Guildford and Havant. An outline study has shown that by operating tilting multiple units on this service the end-to-end journey time of the fast and semi-fast trains can be reduced by nine or ten minutes. Taking this into account, and still allowing for realistic terminal turnround times, then one complete train can be saved.

A reduction in journey time of ten minutes on this route represents a reduction of 10% to 11%. If, to make a hypothetical case, we suppose market elasticity to be of the order of 1.0, then the benefits from introducing tilting trains on the Portsmouth direct line could be:

- one less train in service (14% fleet reduction);
- 10% more passenger revenue (from increased travelling).

Thus, for an increase in procurement expenditure per train of about 10%, the purchaser could procure one less train, and the operator could obtain about 10% more revenue. That <u>must</u> be a good equation!

2.6 Infrastructure

On high speed routes where line speeds are raised at the same time as tilting trains are introduced, there may be once-off costs to ensure that signal spacing matches the higher speed/better braking characteristics of the tilting trains, and also costs (e.g. for level crossing protection and refuges) to upgrade safety facilities on sections where the new line speed profile enters a more restrictive speed range, and to ensure that curve transitions are correctly set and maintained. On routes such as the Portsmouth direct line, where the maximum line speed need not be changed, these costs are likely to be minimal.

3. THE SIMPLE CONCEPT TRAIN

3.1 Simplicity equals better reliability and lower costs

No engineer will deny that reducing the number of reliability-critical components on a train reduces the opportunity for component or train failure. Thus, a train with four traction motors has the potential for experiencing half the number of traction motor failures as would a train with eight motors, all other things being equal.

Also, having fewer components that can fail and which need maintenance reduces the engineering costs of the train.

Other features that can improve train reliability and costs include:

- improving wheel to rail adhesion (reduces wheel tread damage)
- arranging discrete duplication (redundancy) of reliability-critical systems (such as air conditioning)
- having power operated doors away from the body areas that are most affected by load flexing
- having fewer power operated doors for the same entry and egress criteria
- maximising the use of regenerative or rheostatic braking on electric trains (reduces brake pad and disc wear)
- if brake pad wear is a driver of train maintenance intervals, then lower pad wear equals lower train maintenance costs and higher train availability for traffic

3.2 Addressing articulation

British experience in articulation of two carriage body ends on a single bogie has been mixed. The LNER operated large numbers of Gresley carriages with from two to five vehicles on articulated bogies in both main line and suburban applications. The LMS also had a few articulated main line carriages, but did not pursue the idea to volume production.

The BR production advanced passenger train (APT-P) used articulated carriages of full length, ensuring adequate lineside clearance by supporting the vehicle bodies on brackets outrigged from the bogie frame ends. The snag with this arrangement was that vertical and lateral movements of the bogies were accentuated rather than reduced, and were transmitted to the bodies as large and relatively slow vertical and lateral movements. I believe that the resulting ride was a major cause of discomfort and nausea among passengers.

Early European experience of articulation was limited in part to a few diesel high speed multiple units in or near the second world war period. Worthy of mention, however, is a group of articulated diesel multiple unit local train sets that has operated for many years in the Netherlands. In the 1960s and 1970s the railways of East Germany, Poland and Romania put into traffic large fleets of double-deck carriages, many of which were articulated using three-axle bogies to distribute the weight. Also, the modern Danish IC3 diesel trains are articulated.

On the other hand, the numerous tramway systems on the continent employ articulation today as generally standard practice, making great use of its potential for permitting full-width interior gangways between adjacent vehicle sections. Modern light rail or tram vehicles operating on the Tyne and Wear Metro, Docklands Light Railway, Manchester Metrolink and South Yorkshire Supertram systems are also articulated. These operate at relatively slow speeds up to 80km/h, but the bogie/body dynamics are sufficiently good for passengers to be able to ride or stand on the bogie articulation platforms. Some of these systems offer little, or no, cant variation between adjacent car bodies.

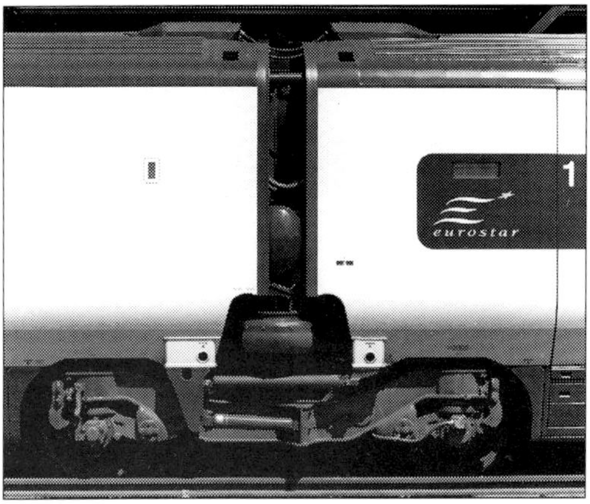

Fig. 3: Articulation on Eurostar high speed train

Fig. 4: Articulation on Manchester Metrolink tramcar

The forms of articulation employed in the single-axle Talgo vehicles in Spain, and in the single-axle, new electric suburban multiple units in Copenhagen, are extreme examples of visionary use of the concept.

Most Railway Division members will by now have ridden on the Eurostar trains and will have experienced the superb riding qualities of these trains. This results partly from the dynamic effects of applying the weight of the two adjacent carriage bodies on the bogie frame as if they were one, that is one vehicle body end is supported by the end of its neighbour.

Thus it appears that modern applications of articulation in the U.K. are well established for very high speed trains and for tramcars, but not for trains in the intervening categories. Where is the logic in this?

3.3 Benefits of articulation

It is useful to keep in front of us the principal benefits that can be attributed to articulation:

- fewer bogies, therefore fewer moving parts, therefore lower maintenance costs
- more space released under vehicles for motors, engines or other equipment
- more space for tilt equipment
- better potential riding qualities
- greater stability in collisions and derailments.

There are also some disbenefits of articulation that must be considered:

- higher axle load (limits maximum weight of vehicle body, and could affect access charges)
- need at depots for bogie or wheel drops (or train-length jacking facility)
- shorter body lengths, requiring more bodies for a given train length.

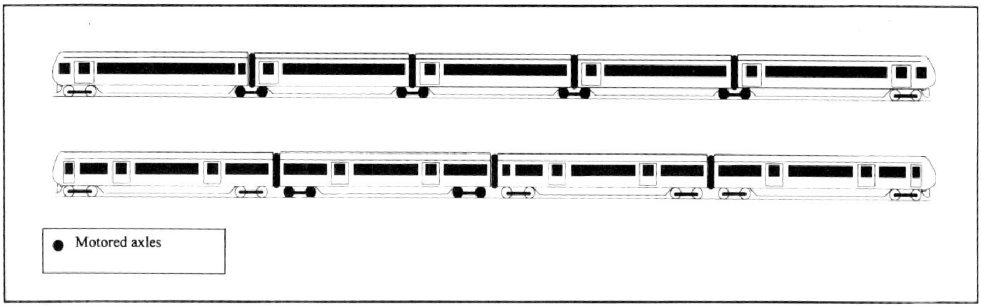

Fig. 5: Comparison between Simple Concept Train (upper) and conventional train (lower)

3.4 Simple Concept Train

This paper offers the Simple Concept Train (SCT) as an improved solution to the design of trains in the most common ranges required for procurement by lessors and operators. Other configurations are discussed in Section 4. As a taster, I offer the following two five-car articulated multiple units:

- a suburban type train of up to 80 metres overall length that can occupy the same track length as a conventional four-car train of 20m vehicles;
- a cross-country type train of 92 to 96 metres length that compares with a train of four conventional 23m vehicles.

These are described in paragraphs 3.6 and 3.7 below.

3.5 Train layout and characteristics

The train layout would have one engine/transmission set or traction motor under each vehicle other than the middle one of the five. The engine/transmission set or motor would drive through a cardan shaft to the nearest wheelset of the adjacent articulated bogie, and the two wheelsets on the bogie would be coupled by cardan shaft, as in many modern diesel multiple units.

Thus four engines or four motors would drive the eight axles of the articulated bogies. This offers about 75% of the train weight for adhesion (compare that with 36% to 57% for conventional suburban electric multiple units). Such high adhesion offers better acceleration and braking from the traction equipment, and a useful amount of regenerative or rheostatic braking. Also, the act of coupling axles that have higher-than-conventional axle loads would

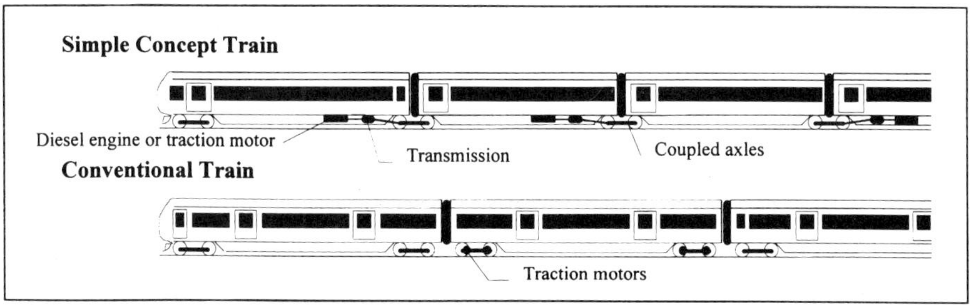

Fig. 6: High adhesion weight from few traction motors

minimise the risk of wheel slip and slide in conditions of low adhesion. Thus the very high costs of the leaf fall season would be much reduced or even eliminated.

The price SCT pays for fewer motors, better adhesion and much lower wheel tread damage is the additional cost of an extra carriage body, and (on the electric multiple unit version) the cost of four additional axle gearboxes and cardan shafts.

There is also potential for simplifying the vehicle body construction. The shorter carriage bodies enable a revision of the traditional passenger access/egress door arrangements. Instead of eight pairs of double-leaf doors on each side of the train at the one-third and two-thirds positions, SCT could be configured with six wider access doorways. These could be situated at the vehicle ends, with wide gangway access from the adjacent vehicles making more than one doorway per vehicle side unnecessary, other than at one extreme end of the unit. Thus the potential is there for simpler body construction using long, extruded sections. The wider doors, and better overall train performance between station stops, enable the SCT to match existing journey times even in typical suburban areas.

Because the access doorways are at the ends, the vehicle seating areas would be uninterrupted by traditional vestibules. This would enable much greater flexibility in seating layout designs, and, for cross-country work at least, a much better and quieter interior environment for passengers. For a given train length, about 15% extra seats could be provided with the same seat spacings as now.

3.6 Suburban SCT

The concept is for a train of five vehicles with bogie centres at the traditional 14.17m distance. The maximum train length with this configuration is about 77m, approaching the 80m of a conventional four-car multiple unit of 20m vehicles.

If account is taken of the greater lateral inertia of two carriage body ends mounted on one bogie, then it appears possible to extend the distance between bogie centres of the suburban SCT to 14.8m, giving a train length of 80m.

A simple concept train of 80m length could accommodate 390 seats in high-density formation instead of about 340 in a conventional train, an increase of nearly 15%.

3.7 Cross-country SCT

The concept is for a train of five vehicles with bogie centres at 17m distance. The maximum train length with this configuration is about 91m, almost the same as a conventional four-car multiple unit of 23m vehicles.

If account is taken of the greater lateral inertia of two carriage body ends mounted on one bogie, then it appears possible to extend the distance between bogie centres of the cross-country SCT to 18m, giving a train length of 96m.

In a simple concept train of 96m length in a cross-country application, 340 seats are possible whereas a conventional train of four 23m carriages would seat about 300. That represents an increase in seating capacity of about 13%.

3.8 SCT weight

The effect of having fewer bogies and shorter, simpler bodies balances the weight of the additional body quite effectively. There need be little overall difference in the weight of a five-car SCT electric multiple unit and a conventional class 319 electric multiple unit. The key difference is an increase in maximum tare axle load from 12.5t (four axles on the power car) to about 15t (the eight driven axles on the SCT). For U.K. operation, there could be a penalty for the higher axle load in the Railtrack access charge, but this would be compensated, at least partly, by the reduced number of axles employed, and by the fact that the train weight is similar to that of a conventional train.

3.9 Cost of new SCT

Assessments of new build costs for conceptual train designs are necessarily conjectural. Nonetheless, it is possible to build up costs based on recent prices of existing train components, beefed up where necessary to cover higher loadings. It appears that a balance can obtain between the effects of having one additional body and two fewer bogies, subject to the potential for reducing body costs through the effect of having fewer passenger door spaces on simplifying body construction. I would guess that a train cost increase of 0% to 7% can easily be sustained by a train with 13% to 15% more seating capacity.

Train cost estimates in a paper such as this must assume that overheads such as the cost of tooling for train construction are spread the same way in each of the types being compared.

Whole life costs would be influenced by the reduced number of maintained key components such as bogies and doors, and by the additional facilities needed at those depots not already equipped with bogie drops or, if really needed, whole train lifting.

4. AN IDEAL UNIVERSAL TRAIN

4.1 Moving forward

Within the U.K. railway industry, the search for a universal train design has so far concentrated on developing existing, proven past products that have given satisfaction in service and which can be produced and maintained at an acceptable price. Thus we see a plethora of new three-car or four-car diesel or electric multiple units with 20m or 23m body lengths, all of which suffer either from having too few driven axles (like the 319), in which case annual adhesion problems are faced, or they have eight driven axles (as in the 365) in which case the number of traction motors is double that needed otherwise to propel the train. None of them to date has offered a simple, add-on tilt package, so for most of them the potential for future, more advanced use of the trains is limited to currently attained levels of performance. Their residual values will ultimately give testimony to this apparent lack of vision.

This paper commends that an ideal universal train would combine the two areas of development offered in this paper:

- active body tilt, on a
- Simple Concept Train.

4.2 Characteristics of an ideal universal train

A manufacturer offering modern trains to diverse customers ought to be able to build on the Simple Concept Train idea in a number of ways. Unit train lengths could be standardised on the 80m and 96m variations for five-car sets, or the components of these used in other formations as required by the customer from three up to, say, eight or ten articulated vehicles. Each formation would be offered in suburban, cross-country or luxury interior appointments to meet customers' needs. Performance would be geared to speed maxima from 120km/h in suburban applications to 200km/h or higher for cross country and main line. Tilt capability would be offered as standard (or as a future additional feature by progressive modification) in the cross-country and main line units.

4.3 Overall benefits

We have seen that the shorter journey times made possible by tilting trains are likely to generate more traffic (in our example 10% seems possible) and enable fleet reductions by earlier turnrounds (in our example, again, a 14% fleet reduction). We have also seen that the Simple Concept Train can provide additional seating capacity more than sufficient to absorb the additional traffic in the same train length, at a first cost that should be little different from a conventional train of similar length. Thus the equation appears to be:

- 10% or more reduction in fleet size
- around 10% more passengers
- 12% to 13% more seats
- broadly similar fleet procurement cost.

I believe this set of numbers points the way forward for the industry.

4.4 Advancing the product range

The paper so far limits the product range of tilting SCTs to suburban and cross-country applications. The product range could be much wider than this. Indeed, I believe that it could embrace virtually the whole U.K. future requirement for railway passenger stock.

Nearer the upper end of the product range, as a replacement for seven- or eight-car HSTs (nine or ten if we count the power cars!) there could be a nine- or ten-vehicle SCT geared for 200km/h with eight or nine underfloor diesel engines each of 375kW output. That is quite modest as a concept. It pushes against no engineering state-of-the-art boundaries, except that the SCT might need to improve slightly on the Danish IC3 train's low transmission of diesel engine noise into the passenger compartment. Longer formations than ten vehicles are, of course, possible. The concept is totally flexible.

Thus, extension of the HST-replacement idea to a fully-fledged 250-plus km/h high speed tilting unit train with distributed traction is but a logical further step forward.

And to think that the SCT was originally conceived as a suburban train!

C547/014/98

The development of gauging software for tilting trains

D M JOHNSON BSc, CEng, MICE, MInstD, FPWI
Laser Rail Limited, Derby, UK
I M GOLDNEY BSc, CEng, MPWI, MIMechE
Interfleet Technology Limited, Derby, UK

SYNOPSIS

Software that is able to compare the kinematic swept envelope of a train with track infrastructure data already enables gauge clearances to be evaluated for conventional trains running on existing tracks. However, before the possibility of introducing tilting trains could become reality in the UK, a new approach was necessary to enable the same process to be undertaken for the more complex evaluations needed for tilting train operation. This challenge brought together Interfleet Technology and Laser Rail and combined their expertise (Interfleet Technology in rail vehicle dynamics and Laser Rail in gauging systems) to develop a method of evaluating route clearances for tilting trains.

Obtaining gauge clearances for trains to run on Railtrack's networks depends on accurate and detailed knowledge of both the train's dynamic performance and of the infrastructure on which it is to run - particularly where limited clearances already exist.

A system has already been developed that compares vehicles' swept envelopes against infrastructure data stored on CD-ROM. This software is now established as the industry standard for kinematic gauging. However, a tilting train's dynamic performance requires more complex analysis and thus cannot be described by a conventional kinematic envelope.

New work was therefore needed to develop a reliable method of assessing gauge clearances for the proposed new generation of tilting passenger trains.

In order to successfully develop the new programme, Interfleet Technology derived a set of generalised algorithms which fully describe a tilt system's characteristics taking into account secondary factors including tilt failure, overshoot, lag and tilt centre movement. With the availability of this new information, Laser Rail was then able to increase the capability of their software to replace a conventional kinematic envelope with the generalised algorithms. It is now possible to compile vehicle specific kinematic data and compare this with known route data to determine clearances.

This paper describes the development of the software and demonstrates how it may be used to assess gauge clearance.

1. GAUGING SOFTWARE DEVELOPMENT

In recent years, the gauging of railway vehicles has developed considerably from the template driven comparison of static vehicle sections with hand-measured structure profiles on 1:20 scale paper plots. Led by the need to allow increasingly larger freight and passenger vehicles through more of the UK rail network, and the wish to accommodate better passenger ride, a combination of gauging techniques and suspension analysis has led to the technique now known as kinematic gauging.

1.1 Kinematic Gauging

The principal conceptual change involved in the progression from static to kinematic gauging is that static gauges must be defined to accommodate all possible movements of the vehicle under any operating condition, whereas kinematic gauging analyses the range of vehicle movements given specific force inputs. Traditionally static gauges only included geometric considerations associated with overthrow on curves. Kinematic gauging, in addition, considers vehicle cant deficiency or excess, speed and a number of variable tolerances. Thus, kinematic gauging presents a more realistic measure of clearances than those produced by static means. In practice, this means that where a more detailed analysis of the vehicle is available, it is possible to allow greater access for that vehicle to the network than would have been possible if only static gauge was considered. This approach is vital in order to clear vehicles with soft suspensions (such as those fitted with air bags). Without such an approach, the inclusion of full suspension movements would result in either an unacceptably large static gauge, or a very restrictive body section.

The modern gauging process consists of measuring the structures, defining the vehicle kinematic envelopes in relation to local conditions and calculating the resulting clearance.

1.2 Structure Data

Structures are measured by a number of means. At the high technology level, it is possible to measure structures at up to 50mph using a vehicle known as the Structure Gauging Train. This vehicle uses cameras to triangulate a white light projection. Whilst fast, its use depends on the availability of train paths, and it must run at night. At the other extreme, manual measurements using traditional platform and structure gauges may be used to measure individual structures. In the middle fall a number of measuring devices, mainly laser based, which are able to measure accurately and swiftly. The choice of a particular method depends on balancing availability, accuracy, speed and cost. All methods have a common form of data output, namely that related to a programme known as Interactive Track Design (ITD) Gauging.

The ITD Gauging programme is essentially a database designed specifically around the needs of structure gauging. In early days, when the software was originally conceived, it was aimed at the static gauging process, and proved very successful at supplementing the techniques used by gauging engineers. The latest systems include full kinematic assessment. The programmes allow structure data to be accessed, and kinematic envelopes of library vehicles to be compared to these and clearances measured.

1.3 Vehicle Data

Vehicles are also collected into a database, which is regularly updated on CD. The library consists of static vehicle profiles, dimensional data affecting geometric performance and suspension characteristic data from which kinematic modelling may be done.

The calculation of kinematic envelopes follows the principles of a former British Railways Design Guide known as Design Guide 501. Vehicle sways and drops are calculated for specific inputs of cant deficiency and speed through a parameter known as equivalent cant. Equivalent cant is a notional parameter which sums the sway forces acting upon a vehicle from both the cant deficiency of excess of a moving vehicle and the forces generated by the lateral accelerations caused by routine track roughness. Equivalent cant is translated into vehicle sways and drops for a variety of suspension conditions, both normal and failed. The computer programme calculates cant deficiency or excess from the known track curvature, installed cant and speed. Tolerances are included to allow for measuring accuracies and track condition. Using lookup tables appropriate to both the vehicle and its condition, the corresponding sways and drops are calculated, and the static envelope of the vehicle adjusted to reflect its attitude. Tolerances for track position and vehicle maintenance status are also included. The addition of curve overthrows finalises the positioning of the envelope within the structure and allows clearances to be measured.

1.4 Clearance Analysis

It is important to note that the analysis must include a number of different scenarios to reflect the range of operating conditions that may occur to a given vehicle. Of particular note is the speed of the vehicle. At maximum speed, the vehicle will sway most to the outside of a curve and this represents the worst outward case. However, it cannot be relied upon that the vehicle will always be travelling at maximum speed. The worst condition to the inside of a curve is known as trundle speed. This is not stationary, but a speed where the sway due to speed alone (simplistically assumed to be linearly related to speed) combines with the inward sway due to lack of centrifugal force to produce a situation where inward sway is maximised. This is typically low on tight curves, but becomes progressively larger as the curve relaxes. On straight track, trundle speed is limited to line speed, resulting in equal sways to each side of the vehicle. Suspension condition must also be considered. Normally, tare vehicles with inflated secondary suspensions result in the highest configuration. Crush loaded vehicles with inflated suspensions result in the worst sway values. Crush loaded vehicles with failed suspensions usually result in a worst case in relation to platform clearances. All appropriate conditions must be analysed and considered in the light of allowable clearances, which differ for normal and failed suspension cases.

Clearances may be considered satisfactory on normal lines if at least 200mm exists above the vehicle solebar and 50mm to platforms. Below these clearances, dispensations need to be obtained, based upon the establishment of suitable risk control measures. Clearance standards are specified in HMRI Railway Principles (the "Blue Book"), and in Railtrack Group Standards.

1.5 Laser Rail Gauging Software

The software to perform these processes is in regular daily use by Railtrack, consultants and contractors involved in the gauging process. Additionally, vehicle builders anxious to develop vehicles to the maximum dimensions possible yet achieving the greatest network penetration are also increasingly using it. Recent additions to the software include the capability of modelling Class 373 Eurostar trains, which need to be gauged to a specific procedure (MSER42). However, this model also uses the basics of Design Guide 501 for its implementation.

2. GAUGING OF TILTING VEHICLES

In developing the new tilt software the intention was to apply the same gauging principles to tilting trains as the ones applied to conventional vehicles as described in Design Guide 501. Design Guide 501 not only gives a series of equations which may be used in analysis of a conventional vehicle, but also describes a process by which gauging calculations are made, such as :-

- calculating vehicle movements at the speed and cant deficiency applicable to a site
- how to allow for dynamic movements (KSpeed)
- how to allow for track tolerances (lateral, cross-level, wheel/rail wear & flange clearance)

Whilst some of these principles could be challenged on the accuracy of their application, the aim of this development was to use the same concepts as used with conventional vehicles, but find a method of predicting the movements of a tilting train within this framework.

2.1 Calculation Method

The objective of the calculation method is to determine the body lateral, vertical and roll displacements for a tilting vehicle with active tilt below the secondary suspension including the following effects, as appropriate :

- body lateral & roll, bogie lateral & roll and tilting bolster degrees of freedom ;
- body, bolster & bogie destabilising stiffnesses ;
- suspension stiffnesses including non-linearities ;
- tilt demand algorithm which may be a function of speed & cant deficiency ;
- effect of track tolerances.

A general tilting train model was set up. The basic calculation method is to obtain a set of co-ordinate displacements for a vehicle model containing non-linear stiffness elements by the derivation of a static force balance with the applied forces on the vehicle components. These results can therefore only be considered to be steady-state conditions. The system force balance is calculated for wide ranges of both speed and cant deficiency resulting in arrays containing displacement results. The displacements of the vehicle are then calculated in terms of lateral, vertical and rotational movements of the vehicle body from the initial position of the vehicle's centre-of-gravity. These are then saved to disk for use in the gauging software.

2.2 "Equivalent Cant"

When analysing a conventional train in order to allow for dynamic movements due to track irregularities at a given train speed, an equivalent cant input is made to the vehicle which gives a static displacement equivalent to the amount of dynamic displacement expected. This is added to the vehicle cant deficiency (or excess) plus an allowance of half the cross-level error to give a total equivalent cant input to the vehicle for a given track site at a given speed.

For a tilting train there are two types of vehicle body movement that must be considered :

- those due to the active tilting mechanism ;
- those due to the passive suspension system in response to vehicle inputs, including tilt.

The active tilt system will respond to the cant deficiency input to the vehicle. It may also be a function of vehicle speed, although usually this is a simple "on-off" function where the tilt system is inoperative below certain speeds. The passive suspension system will react to cant deficiency and speed inputs to the vehicle and its response is also dependent upon the tilt response of the vehicle.

The total vehicle response is therefore separately dependent upon speed and cant deficiency such that a combined "equivalent cant" approach is not feasible. The vehicle kinematic envelope is described by a look-up table of vehicle movement against speed and cant deficiency. Although this results in a more complex set of data to be recorded for the tilting train movements than a conventional vehicle's kinematic envelope drawing, this is not a problem since the data is stored in electronic format and is analysed by computer.

2.3 Allowances for Track Tolerances

The applied forces result principally from cant deficiency (but could also include such additional effects as wind load etc. if desired). These co-ordinate displacements are then used to determine the final location of any defined points on the vehicle body for any given input. Superimposed onto the basic inputs of cant are the effects of track tolerances and vehicle speed. In the case of a passive vehicle, the application of track tolerances and vehicle speed is usually straight forward and results in them both being applied towards the outside of the curve, or both towards the inside of the curve, depending upon whether cant excess or deficiency is under consideration.

However, with a tilting vehicle, it is not possible to determine in advance whether track tolerances should be applied towards the outside or the inside of a curve in order to obtain the maximum envelope due to the rotation of the vehicle body. The calculation is therefore performed for both conditions. Similarly, the effect of K_{speed} may also be acting towards the outside or the inside of a curve resulting in a further two conditions requiring consideration. In order to ensure that the worst case will be found, a total of four cases must be considered :-

- Track tolerances : to the outside of curve ; K_{speed} : to the outside of the curve.
- Track tolerances : to the inside of the curve ; K_{speed} : to the outside of the curve.
- Track tolerances : to the outside of the curve ; K_{speed} : to the inside of the curve.
- Track tolerances : to the inside of the curve ; K_{speed} : to the inside of the curve.

2.4 Other Aspects of Tilt train Gauging

In order to accurately gauge for tilting trains it is important to consider other aspects of a tilting train which will affect its gauge, aspects which are not included or not relevant to a conventional train.

The following aspects are considered :-
- compression drop
- tilt lag, tilt precedence and tilt overshoot
- air suspension deflated

2.4.1 Compression Drop

For a railway vehicle curving on canted track the centrifugal acceleration acts parallel to the ground. Thus, there is a component of the centrifugal force which, if the vehicle is assumed to be in a position normal to the canted track, acts so as to compress the suspension of the railway vehicle. There is normally an additional angle between the vehicle body and the plane of the rails. For a conventional vehicle at cant deficiency the body will roll outwards on a curve which tends to reduce the total angle between the body and the horizontal. For a tilting train tilting towards the inside of the curve this is additive to the installed cant which increases the total angle. Hence, whilst this effect is small for a conventional train it may become significant for a tilting train at high levels of tilt on heavily canted track.

An expression to allow for this effect has been included within the software.

2.4.2 Tilt Lag, Tilt Precedence & Tilt Overshoot

Tilt lag is the difference between the actual tilted position of the train at any instant in time and the ideal tilt position (as required by the tilt algorithm) required for the input to the train. On the entry transition to a curve the vehicle will be tilted less than its ideal position and on the exit transition it will be tilted by more than its ideal position.

Tilt precedence is a method used to try to minimise tilt lag in that each tilted vehicle takes its tilt signal from the vehicle in front. However, any allowance made in the gauging process for tilt precedence and tilt lag must also allow for the leading vehicle which cannot have tilt precedence.

Tilt overshoot occurs at the end of transitions and, unless the system is critically damped, is due to stroke of the tilt actuators going beyond their final position before reaching their final position. There will also be a certain amount of "passive overshoot" due to the passive suspension's response to both the tilt system and the transition.

In making allowances for these factors it is important that due allowance is made for any additional movements that exist due to a tilting train, which exist on conventional vehicles but are considered minor. For instance, there will be a certain amount of overshoot on a vehicle with passive suspension, yet the conventional gauging process ignores this.

Analysis was carried out in order to assess: -

- the significance of the sway movements due to tilt lag and overshoot;
- how to account for these movements within the tilt train gauging process;
- how to calculate the amount of movement to be allowed for a particular type of tilting train.

The findings of this investigation show that it is necessary to make an allowance for tilt lag and overshoot. A passive vehicle will tend to overshoot at the cantrail position (in a lower sway -type mode) towards the outside of the curve at the end of an entry transition. A tilting vehicle in the same situation will tend to overshoot towards the inside of the curve at cantrail level and the outside of the curve at solebar level due to overshoot of the tilt system.

The proposed method to account for tilt lag and overshoot is to carry out a simulation of the actual tilting train. This would include tilt algorithm and active system components, through worst case entry and exit transitions in order to find the maximum differences between the position of the vehicle body including the tilt lag, overshoot effects and the position of the vehicle body that would result from a perfect response. These differences are taken for directions towards the inside and towards the outside of the curve and added to the tilt active results table.

Two tilt active tables will then exist, one with no allowance for tilt lag and one with a maximum allowance for tilt lag. The programme will then be run for gauge clearance for a route in tilt active mode initially with maximum allowance for tilt lag throughout the whole route. This is expected to show 80-90% of structures to be clear using the worst case vehicle gauge. The remainder of structures will then be analysed as to their proximity to transitions and appropriate allowance for tilt lag and overshoot (if any) made for each instance for gauge clearance.

2.4.3 Air Suspension Deflated

For the case of air suspension deflated, since the train is not in its normal operating mode, it is most likely that the tilt system will be "off". Therefore, this case is similar to a conventional train with air suspension deflated.

If the tilt system was to remain active with air suspension deflated, this gives a very complex system. With the air suspension deflated the body rests on friction pads which are fixed to the top of a rubber shear spring. For a conventional vehicle, cant deficiency forces may cause the friction between the shear pads and the vehicle body to be broken out, resulting in the vehicle body being statically offset from the centreline of the track. For a tilting vehicle, the active tilting system and cant deficiency forces will both have an effect on the frictional behaviour.

Since the vehicle body could, in many instances, be in almost any lateral position, dependent upon its initial conditions and subsequent inputs leading up to the analysis position, the method of gauging such a system would be to consider its worst possible position at the point for analysis. A certain amount of work has been carried out to consider solving for such a system, but since it is considered that it is unlikely to occur in practice (i.e., tilt will not be active for air suspension deflated) this area of work has not been completed.

For the deflated suspension cases most likely to be encountered (i.e., tilt will be "off" for air suspension deflated) the kinematic envelope data can be calculated in a similar fashion to the tilt hard failure and tilt passive cases.

3. PROCEDURE FOR GAUGING OF TILTING VEHICLES

In undertaking a gauge clearance calculation for a tilting train it is necessary to take account of the tilting train operating in all load conditions and in all modes of operation. For a conventional train it is usual to undertake gauge clearance calculations for load conditions of tare and crush, and for air suspension inflated and deflated. For an active tilting train it is also necessary to take account of different modes of operation of the tilting system, namely:

- tilt active, i.e. the normal mode of operation of the tilting train ;
- tilt passive, i.e. the tilt system is "off", either by intention or following failure of the tilting system ;
- tilt failed, i.e. a failure of the tilting system which does not result in the tilt system going to tilt passive mode, of which the most extreme example must be a "hard" tilt failure whereby the vehicle body remains in its fully tilted position.

The train maximum speed relevant to each case must also be considered. Further modifications to the programme have been performed to allow intended speed profiles to be used in analysis, rather than restricting speed to the current line speed, since the purpose of tilting trains is to allow operating speeds higher than current line speed. For a failed trainset, gauging may be required to be undertaken at the higher tilt active speed, or at reduced speed, depending upon whether failure has just occurred or whether the train set is continuing its journey following failure. The required gauge clearance may be different for each case. A risk assessment needs to be performed to determine satisfactory operating conditions and clearances for tilt failed trainsets.

4. CONCLUSIONS

The gauging software used for gauge clearance of conventional trains has been established for around 5 years. Whilst not yet complete, a substantial part of the UK network has been measured and placed in various databases. The process of measuring outstanding structures is ongoing. Structures are validated for integrity and placed on the database. They are then cross-referenced to a geographical information system, which relates their location to a traditional railway line diagram. This knowledge enables vehicles to be designed to the maximum dimensions possible, yet achieves the greatest possible network penetration.

The addition of the tilting train capability to the gauging software provided an interesting challenge to Laser Rail and Interfleet Technology and has resulted in a gauging package of benefit to potential manufacturers and operators of tilting trains and infrastructure controllers. The gauging process has been independently checked and verified and was done with approval of Railtrack.

The inclusion of tilting train analysis in the software has enabled train builders bidding for current contracts to consider infrastructure implications at an early stage in the process, thus allowing them to maximise the potential vehicle size with respect to the available space.

The software is now being further developed to accommodate principles known as "Route Opportunity", to provide a programme more orientated towards full route clearance exercises and the provision of a digital route compendium, describing where rolling stock may run with acceptable clearances. This development, based upon a networked Windows NT platform, may be described as "power gauging", since it enables assessments to be performed more rapidly than has even been possible in the past.

5. References

1 Railway Group Standards, Various, Railtrack Safety and Standards Directive.
2 Design Guide 501, British Railways Board, Dept. of M&EE, October 1985.
3 Mechanical Systems Engineer Report No. 42, European Passenger Services, February 1995.

Laser rail

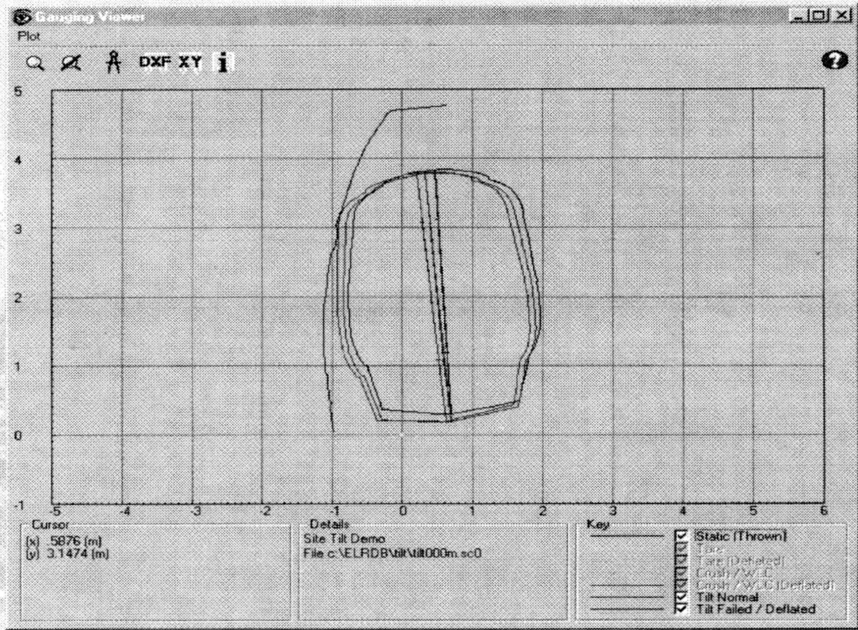

Tilting train travelling at 35 kph

ClearRoute™

Laser rail

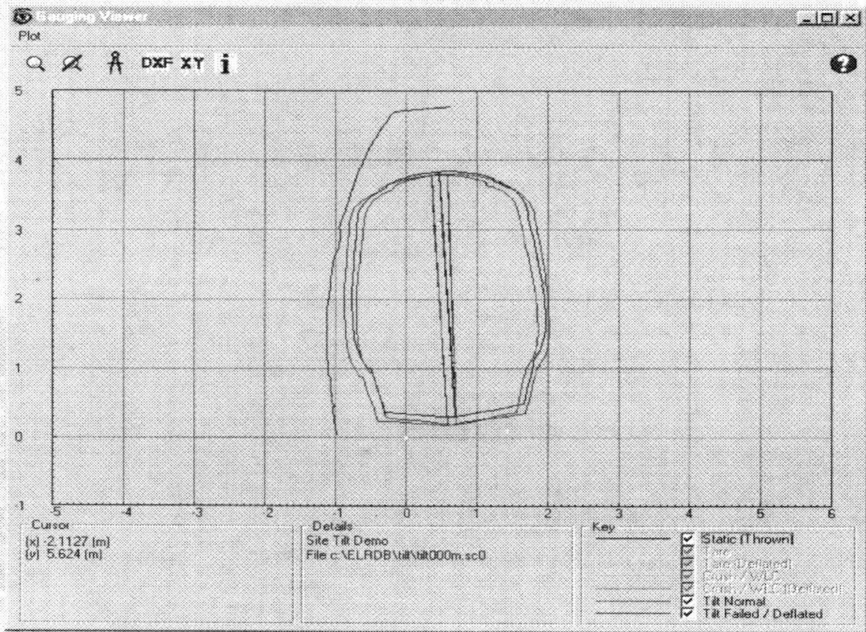

Tilting train travelling at 85kph

ClearRoute™

Laser rail

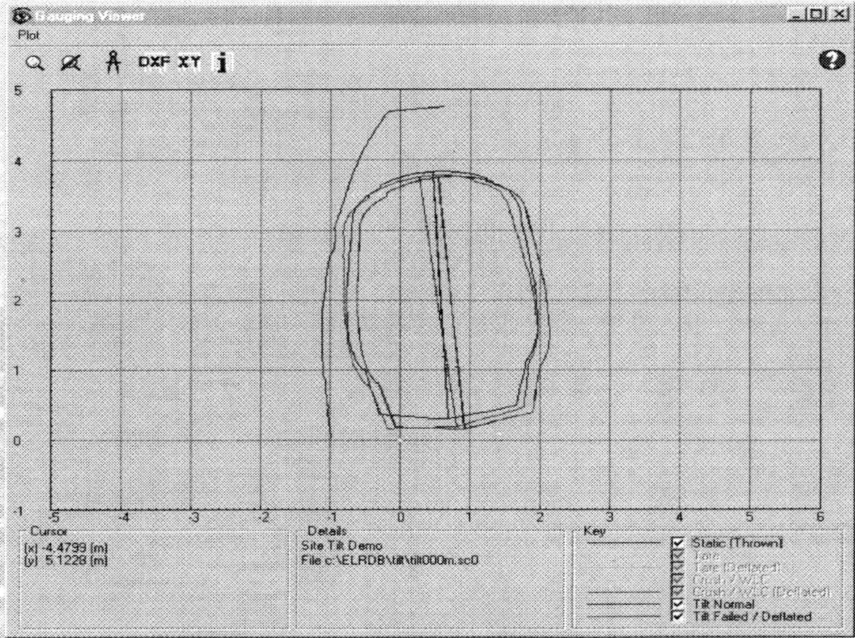

Tilting Train travelling at 130 kph

ClearRoute™

Laser rail

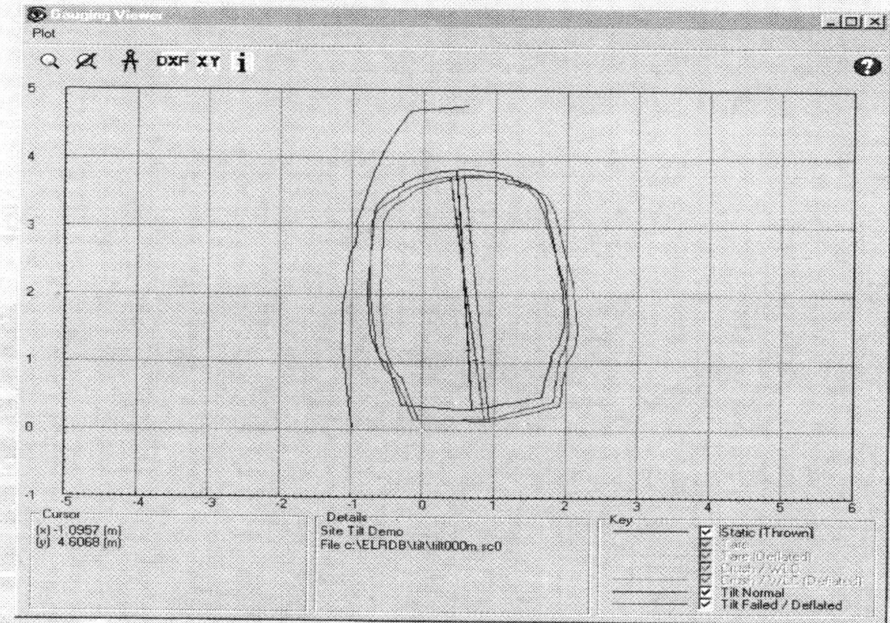

Tilting train travelling at 225 kph

ClearRoute™

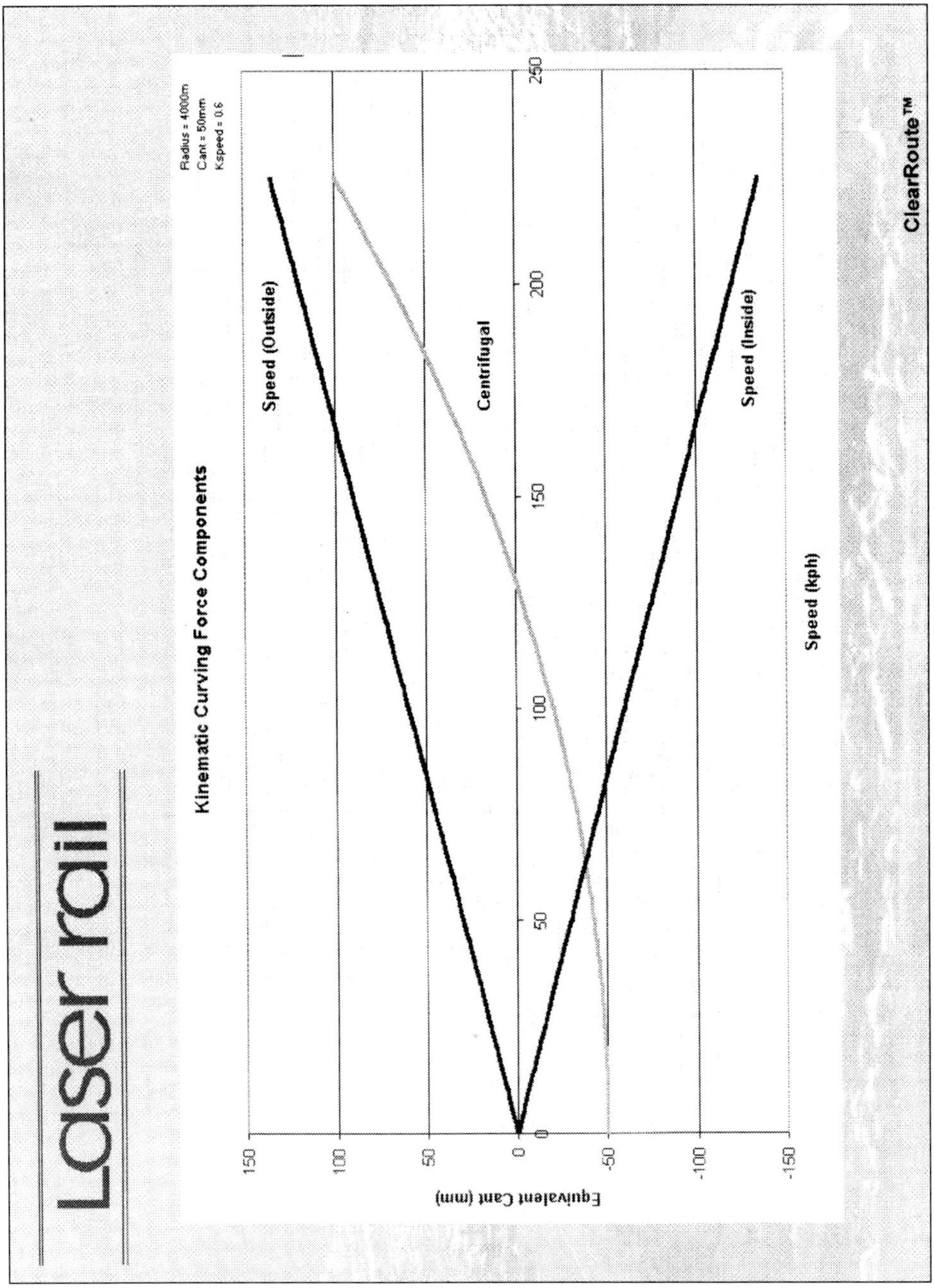
Laser rail
Kinematic Curving Force Components
Radius = 4000m
Cant = 50mm
Kspeed = 0.6
Speed (Outside)
Centrifugal
Speed (Inside)
Equivalent Cant (mm)
150
100
50
0
-50
-100
-150
0
50
100
150
200
250
Speed (kph)
ClearRoute™

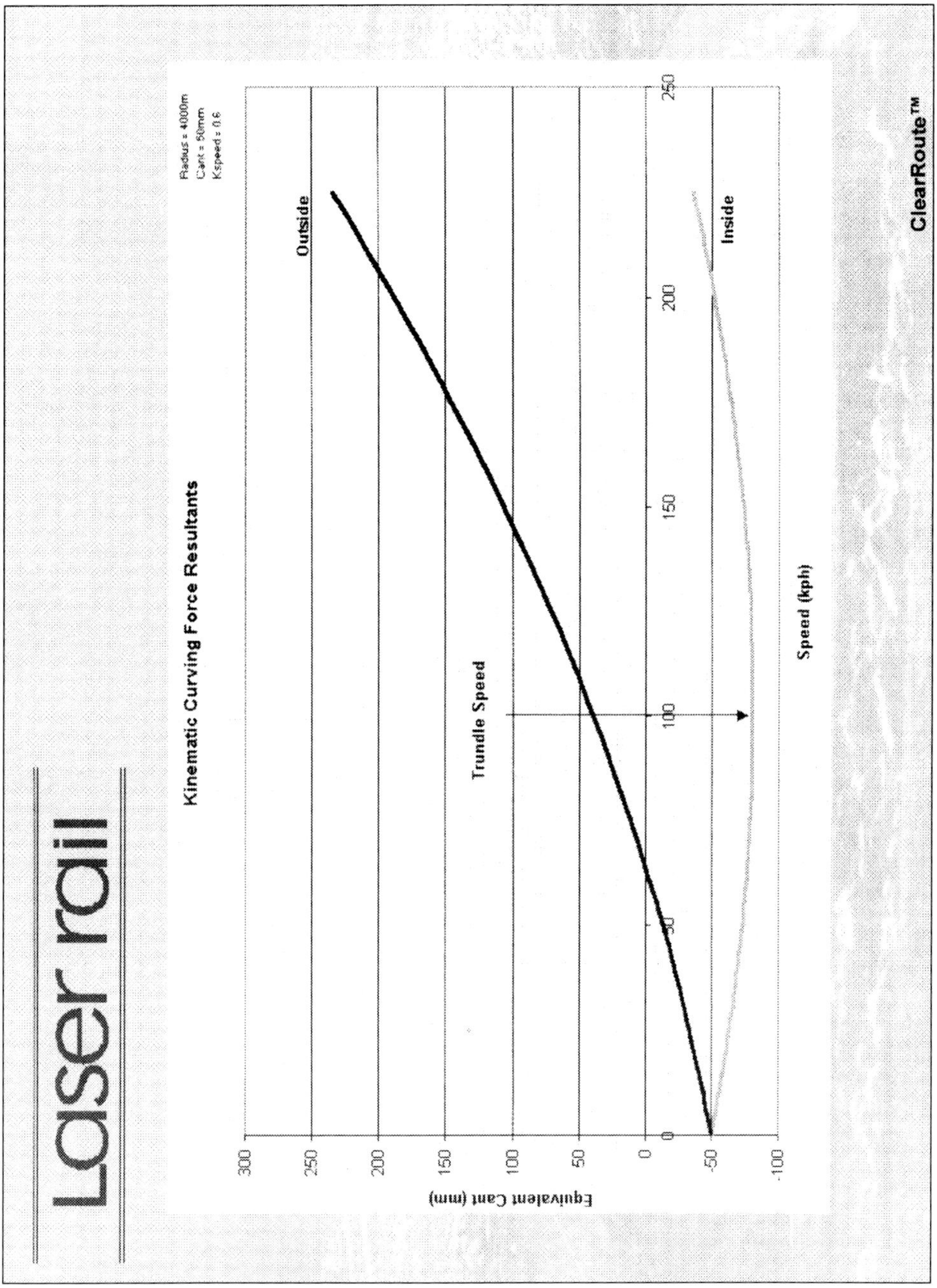
Laser rail
Kinematic Curving Force Resultants
Radius = 4000m
Cant = 50mm
Kspeed = 0.6
Outside
Inside
Trundle Speed
Equivalent Cant (mm)
300
250
200
150
100
50
0
-50
-100
0
50
100
150
200
250
Speed (kph)
ClearRoute™

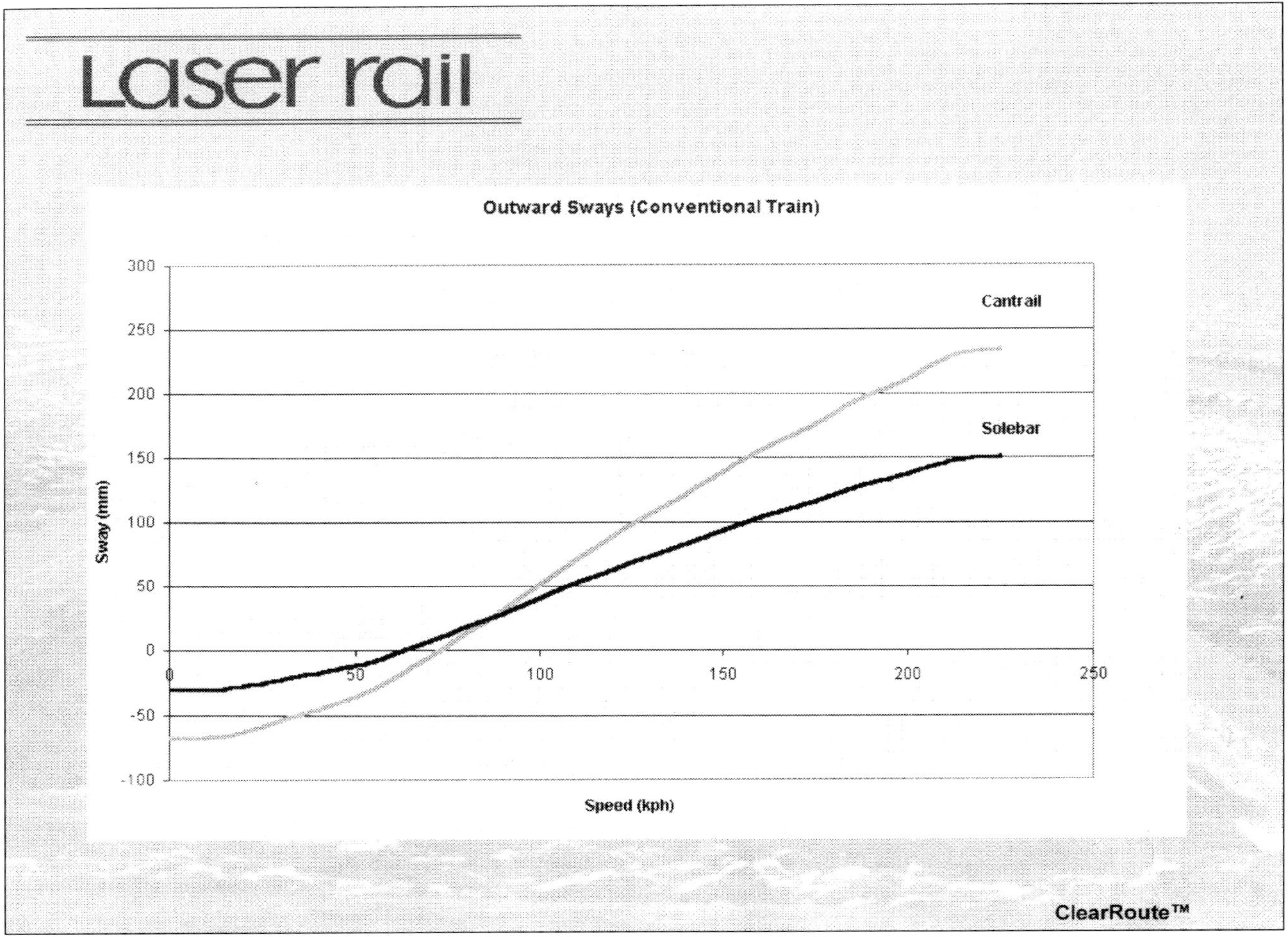
Laser rail
Outward Sways (Conventional Train)
Sway (mm)
300
250
200
150
100
50
0
-50
-100
Speed (kph)
0
50
100
150
200
250
Cantrail
Solebar
ClearRoute™

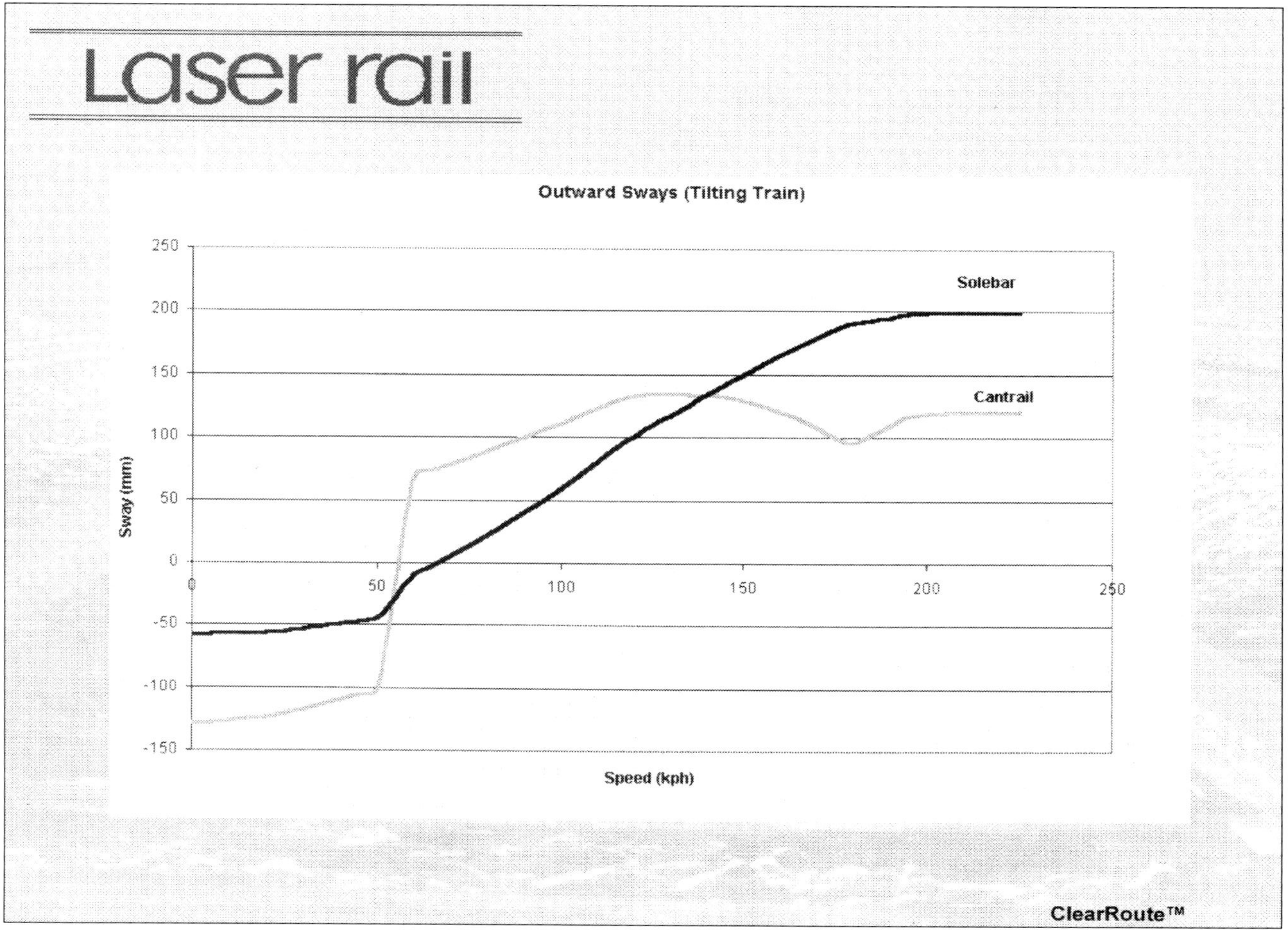
Laser rail
Outward Sways (Tilting Train)
Solebar
Cantrail
Sway (mm)
Speed (kph)
250
200
150
100
50
0
-50
-100
-150
0
50
100
150
200
250
ClearRoute™

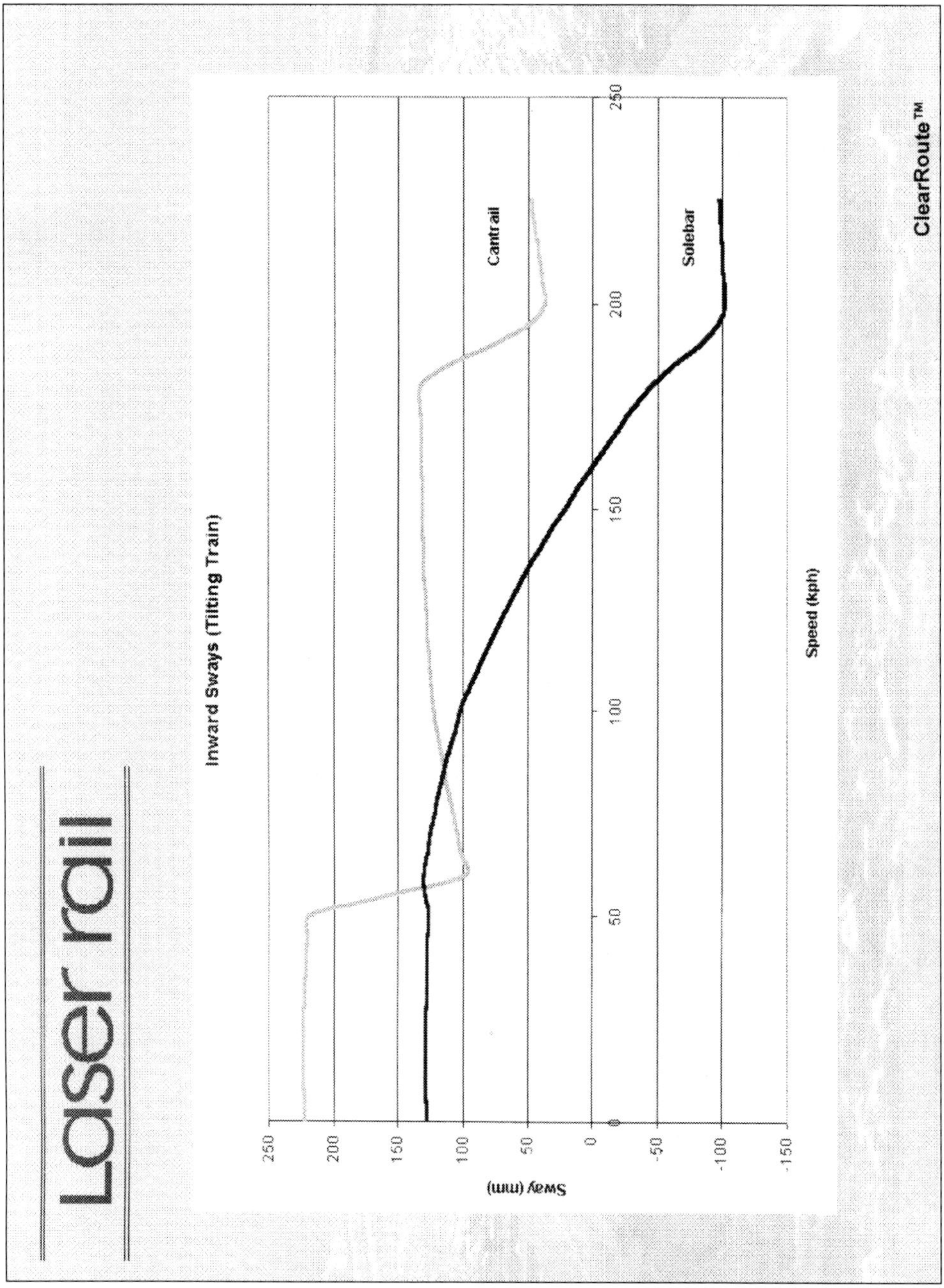
Laser rail
Inward Sways (Tilting Train)
Cantrail
Solebar
Sway (mm)
250
200
150
100
50
0
-50
-100
-150
50
100
150
200
250
Speed (kph)
ClearRoute™

C547/018/98

Innovation and new trains

D SHERRIN
Northwest Trains Limited, UK

SYNOPSIS

This paper outlines the principles and methods adopted to introduce a high level of effective innovation into new train projects. It uses the project for the procurement and construction of thirty new class 458 electric multiple units to illustrate the areas where innovation can be utilised, namely: innovation in design, innovation in management processes and cultural innovation.

1. INTRODUCTION

GEC-Alsthom (GEC-A) is building a fleet of thirty new four car electric multiple units, designated class 458 units, for Porterbrook Leasing Company. Once commissioned the units will be leased to South West Trains (SWT) for use on outer suburban services radiating from London's Waterloo Station. The units represent one of the first fleets of electric multiple units to be procured and introduced within the framework of the UK's privatised railway industry. The successful introduction of these units is critical to SWT, Porterbrook and GEC-A. However, incremental improvement over previous rolling stock projects is not enough. A step change in performance is required. The units MUST be accepted into passenger services on time, to budget, and perform as advertised. To do this requires both careful management and a high level of innovation.

But what is Innovation?

2. OUTLINE DESCRIPTION OF UNIT

Before going further a brief description of the unit will help illustrate some of the arguments in the paper. The unit is a derivative of GEC's Juniper Modular Electric Multiple Unit design. The unit has the following features:-

1. Independent cab and saloon air conditioning
2. Advanced insulated gate bi-polar transistor technology (IGBT) three phase a.c. propulsion equipment with regenerative braking.
3. State of the art bogies with pneumatic secondary suspension.
4. An advanced steel bodyshell design with enhanced crashworthiness to the latest Group Standards.
5. Driving End Gangways.
6. Maximum speed of 100 mph.
7. Compatible with the 750volt electrified third rail network.

So, where is the innovation within the unit? Why and how have we introduced innovation?

3. EFFECTIVE INNOVATION IN DESIGN

I define innovation as "the successful exploitation of new ideas". As SWT Project Manager for the units, and an engineer by profession, I have tried not to talk up the benefits of all singing all dancing systems in the name of innovation. This course of action only leads to escalating capital cost and increased complexity. The units only feature new and novel systems which SWT considers critical to the project's success. SWT has tried not to mistake technical advancement for innovation but to keep to tried a tested practices unless there is a clear benefit.

SWT has used a combination of experience and analytical tools to ensure the introduction of any innovation shows a clear benefit. This is important as innovative designs of equipment have the potential to increase the costs and carry additional risk. In short, innovation must be effective.

3.1 Critical Success Factors

The process adopted for ensuring effective innovation follows a few simple steps. The first step is to identify the critical success factors for the project. These are the minimum number of requirements that must be satisfied to allow the project to succeed. The process of identifying the critical success factors starts with a review of the company's Business Plan. Building on this is a review of the Business Case for the project. From this review, six critical success factors were identified:-

Money:-	The business case for the introduction of the units is very tight. Therefore any additional costs have to be clearly justified.
Time:-	With only a seven year franchise there could be no delay in the introduction of the units.

Reliability:- The units will operate on suburban routes where the impact of any delays will be the most damaging to SWT, therefore the units have to be reliable.

Safety:- SWT has a responsibility to ensure that the units improve the level of safety.

Commercial:- SWT is not looking for gimmicks, just good, well-thought-out design. Litter bins have to work and seats have to have enough leg-room.

Franchise:- The introduction of the units must not put the Franchise Agreement or the licence issued by the Rail Regulator at risk.

3.2 Key Contributing Factors

These critical success factors were then broken down further to identify the key contributing factors. These key contributing factors represented areas where innovation was required to achieve a step change in performance.

For example, the specification and procurement contract was quite clear in apportioning the responsibility for the unit's reliability to the designer, GEC-A. However SWT identified doors and propulsion equipment as key contributing factors to the reliability of the unit. Thus these two areas attracted closer scrutiny to ensure a step change in performance over existing equipment. Joint discussions were held between SWT, Porterbrook and GEC-A to formulate creative ideas for improving the performance of these two systems. The result was the development of the unit's innovative power arrangement and introduction of the Train Management System to improve door performance.

3.3 Examples of Innovative Design

The unit's propulsion system is designed to provide a step change in unit reliability. The unit consists of the following vehicles:-

DMOC:- Driving Motor Open Composite
(P)TOSL:- Trailer Open Standard Lavatory (Pantograph optional)
MOS:- Motor Open Standard
DMOC:- Driving Motor Open Composite

The arrangement of powered and un-powered wheels is shown in figure 1 (the black dots indicate powered axles).

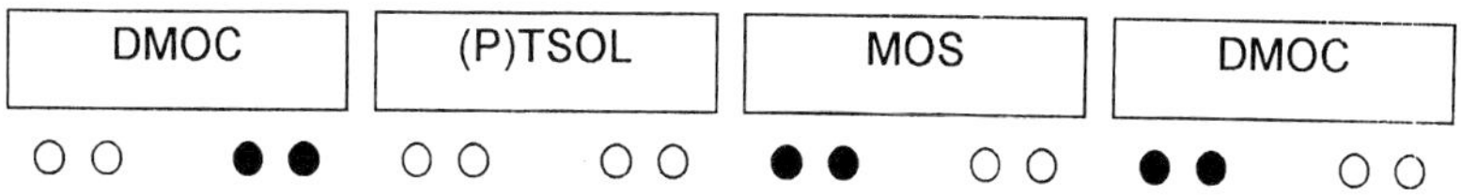

Figure 1-Arrangement of Powered Axles.

Each four car unit has three motor coaches, each with two powered axles. Normally the motor coaches operate at two thirds of their nominal rating. However if one motor coach fails the remaining two motor coaches automatically step up to full power. Thus the unit can remain in service without causing delay. The six powered axles also prove useful in conditions of poor adhesion.

The door system is interfaced with the unit's Train Management System (TMS). The TMS consists of a data bus running the length of the train, which can be formed of several units. The data bus is interfaced with key train subsystems, including individual pairs of doors, and allows Traincrew to monitor and control sub-systems via a console in the driving cab. Using the TMS, Traincrew can locate a defective pair of doors and take the appropriate action. The TMS also stores faults and performance information which can be retrieved by maintenance staff and used to help diagnose defects and persistent service problems.

4. EFFECTIVE INNOVATION IN MANAGEMENT PROCESSES

The discussion so far has focused on the introduction of innovation in the form of novel systems and designs of equipment. However, innovation is not just about research and development or science and technology. Whilst these are important aspects of innovation, innovation itself involves attitudes and practices stretching over a much wider range of activities. Therefore, if the unit is to be truly innovative, we will need to look at more than just the design of the doors and propulsion systems. We will also need to look at the management processes.

It is important when investing in innovative designs that there is a means of evaluating the effectiveness of the innovation. One tool used by SWT to quantify the effectiveness of innovation and to evaluate any proposed improvements was based on simple quantified risk assessment techniques. The first step is to identify an activity, system or process, to evaluate it in terms of creativity and structure, then to apply a notional rating to the level of creativity and structure. Figure 2 shows a simple graph on which the values for structure and creativity can be plotted. The four quadrants of the graph have been given titles which indicate the effectiveness of innovation in each quadrant. The objective is always to have systems in the top right hand quadrant.

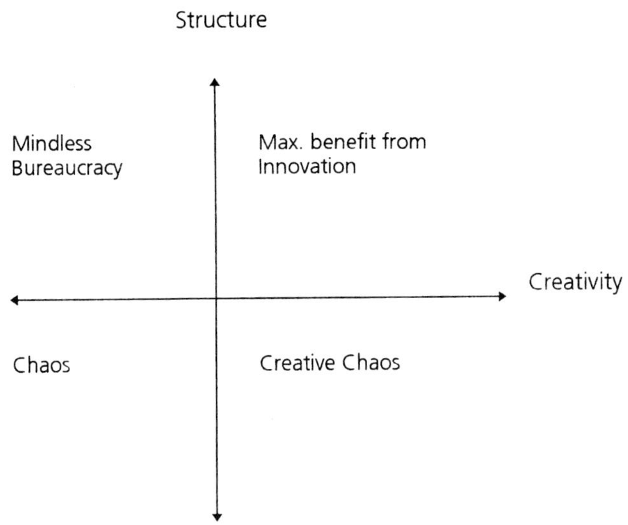

Figure 2 -Rating of Innovation

The tool has been useful in evaluating management processes. Often, extended supplier chains have effectively increased the structure in a process, and stifled creative ideas, resulting in slow bureaucratic processes. The obvious solution is to pull all the parties together and breakdown contractual barriers, however this can end in chaos as everyone throws in ideas, some of which may not be based on a clear understanding of the problem. What is required is to bring parties together, but still keep a structure to it.

4.1 Creative Chaos

Standards are also important in defining structure. A lack of clear, well-written standards can result in chaos. An example of this is the non-compatibility of the new designs of electric multiple unit being produced by different manufacturers. Each manufacturer is introducing novel and innovative systems, but each uses a different design of electrical unit end coupling. This restricts the flexibility to Operators and Leasing Companies.

4.2 Examples of Innovation in Management Systems

The units were procured by competitive tender against a performance specification. The specification and its associated contractual framework represented the first opportunity to introduce a foundation of innovation into the project. Over the years British Rail used a variety of methods, each with varying degrees of success. The culmination of British Rail's procurement strategy was in the late 1980's, where manufacturers were asked to tender against weighty performance specifications which defined the performance characteristics of every conceivable sub-system and component. This was effectively design by specification, and effectively dictated the design adopted by the manufacturer. The result was often technically-similar designs with price being the only means of differentiating between manufacturers' offers.

Porterbrook and SWT adopted a different approach for the specification of the units. Part of the philosophy was to provide a structure through which manufacturers could propose creative ideas and solutions. The specification was, in the main, very subjective. It set out SWTs' general requirements for the unit. The specification was, however, objective in certain aspects to ensure there was a clear definition and allocation of the risk. These areas were limited to areas critical to the success of the project. This approach allowed GEC-A to offer new ideas which it had successfully developed with other Operators. For example GEC-A introduced "Selective Door Re-close" as a feature of the door control system. Selective Door Re-close allows the Conductor to open and re-close an individual pair of doors that has failed to close correctly. This feature will prove very useful at busy commuter stations where passengers can prevent doors closing.

Underpinning the specification is a set of contractual agreements which provides a clear, concise framework within which risk is apportioned to the different parties. In apportioning risk it is important that risk is allocated to those parties who are competent to manage the risk. For example, GEC-A does not bear the risk for interfacing with Railtrack on operational issues. This is clearly SWTs' area of expertise. Conversely, GEC-A bears the risk for gaining Railtrack approval for the electromagnetic compatibility of the unit. This also gives each party the freedom to innovate.

Much has been written about the Railtrack Acceptance process. Clearly this area required an innovative approach which would ensure timely approval of the unit. When looking at this area it was clear that there was no shortage of creative ideas within GEC-A as to how to overcome some of the technical challenges. What was missing was a structure to exploit these ideas. Whilst the introduction of the Rolling Stock Acceptance Board made a significant step forward to improve the structure of the process, both GEC-A and SWT felt that more action was required. Thus a joint steering group was formed. Membership included representatives from SWT, Railtrack HQ, Railtrack Zone, Metro Cammell and GEC-A Traction. The group did not focus on the technical issues, but concentrated on the process, ensuring everyone knew what was going on and what the issues were.

These examples clearly demonstrate the importance of innovative management processes that encourage and allow innovative equipment design to be exploited. However the exploitation will not be effective without an effective culture of innovation within the project. Innovation in people is required.

5. CULTURAL INNOVATION

If "structure and process" are the skeleton and nervous system of a project, then "people" are the blood and guts. Without understanding and mobilising the innovative energies of the people, within a project no effective innovation can be introduced, regardless of the management processes and technical expertise involved in the project.

Implementing a strategy which introduces a culture of innovation into a project cannot happen over night. The pace with which a culture of innovation can be introduced is very dependent on the culture of the organisations involved. However, SWT has put in place a number of principles which will encourage people to seek innovative solutions to problems and create a step change in performance.

5.1 Communicating the Need for Innovation to Sub Contractors

The first principle is to keep teams small and lines of communication short and simple. Following on from this is a policy of getting out and about at all levels of the supply chain and telling suppliers about SWTs' business and its problems. This has created an environment which encourages constructive communication between the user, the designer and the manufacturer. An example of this policy is the relationship which has developed between SWT, GEC-A and Transintech. Transintech is responsible for supplying many of the units' interior fittings, including the seats and body-side panelling. Transintech and SWT have joined together to trial a number of seats in-service. This trial has generated discussion and debate surrounding existing problems, and allowed Transintech to put forward novel solutions. This has been achieved without compromising the contractual relationship between GEC-A and Transintech.

5.2 Communicating the Need for Innovation to Staff

There is also a requirement to communicate the need for change, the need to innovate within our own organisation. The project team has ensured that the new units are a regular feature in our in-house company magazine. A full size mock-up of the unit was displayed at Waterloo, and provided a catalyst for staff and passengers to suggest ideas. SWT has also interviewed a cross-section of staff and asked them where they think problems will arise.

Further on in the project, SWT and GEC-A will be organising a series of visits for SWT Traincrew and maintenance staff to see the units under construction. We are also organising visits for GEC-A assembly-line staff to SWT depots and stations to see how SWT operates and illustrate how important it is that the units work first time. With delivery of the first unit SWT will be seconding staff into the GEC-A commissioning teams. This will enable SWT staff to learn about the units first-hand from day one. It also means that when the commissioning teams leave, not all of their knowledge leaves with them.

6. CONCLUSION

So in summary GEC-A, Porterbrook and SWT have worked together to produce a variety of "innovative" improvements to the class 458 project. We intend these to produce a step change in performance over previous projects. The introduction of innovation has focused on three areas:- innovation of management systems, innovation in design, and introduction of a culture which promotes and encourages innovation. To do this, we have used a combination of experience and analytical tools to identify where innovation is needed and how to ensure it is effective. Finally, whilst innovation in design and innovation in management processes can be developed independently, both are reliant on all the people in the project having an innovative culture. Figure 3 illustrates this dependency.

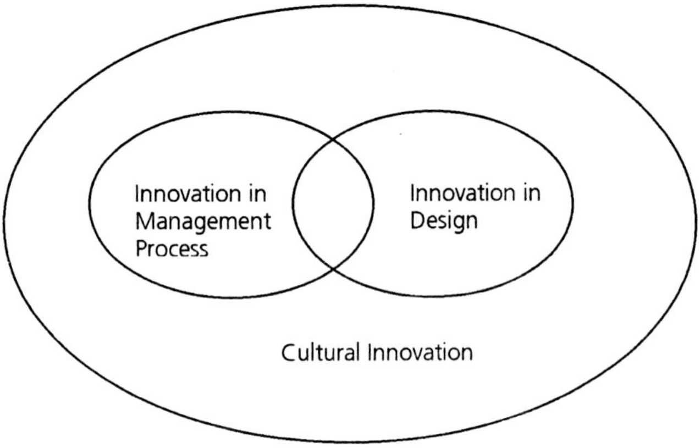

Figure 3 -Dependency of Innovation

Authors' Index